Natural Resources and the Environment Series
Volume 6

Renewable Sources of Energy and the Environment

NATURAL RESOURCES AND THE ENVIRONMENT Series

Volume 1: THE ENVIRONMENTAL IMPACTS OF PRODUCTION AND USE OF ENERGY
Essam El-Hinnawi, United Nations Environment Programme.
ISBN 0 907567 00 2 Hardback, 0 907567 15 0 Flexicover.

Volume 2: RENEWABLE NATURAL RESOURCES AND THE ENVIRONMENT: PRESSING PROBLEMS IN THE DEVELOPING WORLD
Kenneth Ruddle and Walther Manshard, United Nations University.
ISBN 0 907567 01 0 Hardback, 0 907567 06 1 Flexicover.

Volume 3: ASSESSING TROPICAL FOREST LANDS: THEIR SUITABILITY FOR SUSTAINABLE USES
Richard A. Carpenter, Editor, East West Center.
ISBN 0 907567 02 9 Hardback, 0 907567 07 X Flexicover.

Volume 4: FUELWOOD AND RURAL ENERGY PRODUCTION AND SUPPLY IN THE HUMID TROPICS
R. P. Moss and W. B. Morgan, United Nations University.
ISBN 0 907567 03 7 Hardback, 0 907567 08 8 Flexicover.

Volume 5: ECONOMIC APPROACHES TO NATURAL RESOURCE AND ENVIRONMENTAL QUALITY ANALYSIS
Maynard M. Hufschmidt and Eric L. Hyman, Editors, East-West Center.
ISBN 0 907567 04 5 Hardback, 0 907567 09 6 Flexicover.

Volume 6: RENEWABLE SOURCES OF ENERGY AND THE ENVIRONMENT
Essam El-Hinnawi and Asit K. Biswas, Editors.
ISBN 0 907567 05 3 Hardback, 0 907567 10 X Flexicover.

Volume 7: GLOBAL ENVIRONMENTAL ISSUES
Essam El-Hinnawi, Editor, United Nations Environment Programme.
ISBN 0 907567 11 8 Hardback, 0 907567 12 6 Flexicover.

Volume 8: THE STATE OF THE GLOBAL ENVIRONMENT: A DECADE AFTER STOCKHOLM
United Nations Environment Programme.
ISBN 0 907567 13 4 Hardback, 0 907567 14 2 Flexicover.

Renewable Sources of Energy and the Environment

ESSAM EL-HINNAWI *and* ASIT K. BISWAS
Editors

Published by TYCOOLY INTERNATIONAL PUBLISHING LTD.
DUBLIN

Tycooly International Publishing Ltd.
17 Gilford Road
Sandymount
Dublin 4, Ireland
Telex 31539 TVES EI

First edition 1981.

Typeset by Brunswick Press Ltd., Dublin and
printed by Irish Elsevier Printers Ltd., Shannon, Ireland.

ISBN 0 907567 05 3 Hardback.
ISBN 0 907567 10 X Flexicover.

CONTENTS

FIGURES

TABLES

List of Contributors

Editors

Asit K. Biswas, Director, Biswas and Associates, Oxford, England and Ottawa, Canada.

Essam El-Hinnawi, Chairman, Energy Task Force, United Nations Environment Programme, Nairobi, Kenya.

Contributors

H. Jack Allison, National Science Foundation, Washington, U.S.A.

A. J. Ellis, Department of Scientific and Industrial Research, Wellington, New Zealand.

Fatma El-Gohary, National Research Centre, Dokki, Cairo, Egypt.

Jean Pheline, Commissariat à l'Energie Atomique, Paris, France.

M. I. Said Ahmad, National Research Centre, Dokki, Cairo, Egypt.

Bent Sørensen, Roskilde University Centre, Roskilde, Denmark.

Preface

THE 1970s brought into focus the general realization that fossil fuel resources, especially oil and natural gas, are finite in nature and that countries should explore the possibilities of using other sources of energy as well, establishing thereby an appropriate energy mix to meet their demands for sustainable development. Several scenarios have been proposed in recent years for future world energy demand and supply, but these scenarios are roughly indicative and subject to a wide range of uncertainties because of the different and partly unpredictable assumptions on which they are based. Future energy availability will be influenced by a variety of factors, among which are new discoveries, improved production techniques, development of new energy sources, and economic, social, geo-political and environmental constraints.

The general realization of the finite nature of fossil fuel resources has caused a re-examination of the possibility of using those energy resources which are of a non-depleting nature, and are therefore considered renewable. On a global basis the resource base of renewable sources of energy is extremely large with an infinite potential for renewal. With the present state of technology, it is difficult to estimate how much of the resource base can be technically and economically exploited, and what would be the contribution of renewable sources of energy to the overall future world energy supply systems. However, the potential and role of renewable sources of energy in meeting part of the future energy demand should not be underestimated. In the developed countries strategies for the exploitation of renewable sources of energy constitute a part of present policies which aim at reducing the dependence of fossil fuels. In the developing countries, particularly those short of fossil fuel resources, the renewable sources of energy show promise of meeting some of the future energy needs, especially of rural areas. Without strong rural development programmes based on decentralized energy sources, urban migration will become even more acute, exacerbating the already dire urban problems brought into focus at the United Nations Conference on human settlements in 1976. Similarly, the United Nations Conference on desertification convened in Nairobi in 1977 emphasized the importance of using locally available renewable sources of energy to reduce excessive wood cutting in arid and semi-arid areas, which is one of the most important causes of desertification. The importance of renewable sources of energy has been recently brought into focus by the decision of the General Assembly of the United Nations to convene a United Nations Conference on New and Renewable Sources of Energy in Nairobi in August 1981.

Like any other sources of energy, renewable resources may have a host of interrelated environmental effects within nations, within regions, or on an even wider international scale. Among these, questions of the availability and allocation of resources are likely to play as important a role as the pollution problems proper. Land and water use, emissions and their impact on ecosystems and human health are but examples of the problems encountered. All the effects of energy resources on the development of human life, and on ecosystems and environmental processes generally, need to be better understood, and the long-term real costs and benefits of such environmental alterations carefully weighed. Good management of the environment should be based upon avoiding wastage of resources as well as pollution. This is more appropriate and certainly more efficient than redressing environmental degradation after it occurs.

The present volume, written in collaboration with a number of distinguished colleagues, aims at highlighting the environmental aspects of different renewable sources of energy. The views expressed in the different chapters of this book are those of the different authors and do not necessarily represent those of the organizations and institutions with which they are affiliated or of the United Nations Environment Programme.

We hope that this book will be found to give a balanced assessment of the environmental impacts of different renewable sources of energy and that the scientific community will pick up the many inadequacies in our knowledge of these environmental impacts, and accelerate the efforts to find adequate solutions to the problems encountered or likely to be encountered.

We would like to express our gratitude to our colleagues who contributed to this volume for their whole-hearted co-operation and to Dr. Mostafa Kamal Tolba, the Executive Director of the United Nations Environment Programme, without whose support and encouragement this book could never have been written. Several scientists kindly reviewed some of the chapters and/or provided up-to-date information. To all of them, we would like to express our appreciation. Mrs P. Ayew and Mrs A. Fernandes were responsible for typing the manuscript; their efforts are gratefully acknowledged.

Essam El-Hinnawi
Nairobi, *July 1981*

Asit K. Biswas
Oxford, *July 1981*

CHAPTER 1

The Promise of Renewable Sources of Energy

ESSAM EL-HINNAWI
United Nations Environment Programme, Nairobi, Kenya

Fig. 1. A Karri tree being felled in Australia (photo: FAO).

INTRODUCTION

Since prehistoric times human society has been consuming a constantly increasing amount of energy. Statistical analysis of data for the period between 1925 and 1950 shows that the average annual percentage rate of change in total commercial energy consumption was 2.2 per cent. But in the following ten years, from 1950 to 1960, the figure more than doubled (4.9%), and in the decade after that it reached 5.6 per cent. The total commercial energy consumption in the world in 1950 was about 2500 million metric tons of coal equivalent (tce), in 1970 it was 6,500 million tce and it reached about 8,700 million tce in 1979 (United Nations, 1980). This increase in energy consumption is a natural result of increase in population and of economic development. In 1950, the world population was 2,513 million; the average per capita commercial energy consumption was about 1,000 kg coal equivalent and in 1979 there were 4.300 million people and the average per capita energy consumption was about 2,000 kg coal equivalent. In other words, in about 30 years the average per capita commercial energy consumption in the world has doubled.

It has generally been assumed that there is a consistent positive relationship between gross national product (GNP) and energy consumption. Thus, as a country's GNP in real terms rises over time, its energy consumption goes up as well. However, the degree of dependence, economy versus energy, can be influenced by a variety of factors. For example, the structure of the economy will affect the correlation between GNP and energy consumption; the rate of increase of GNP and energy use is somewhat similar for energy-intensive exports, but for non-energy-intensive exports, GNP increases at a much faster rate than energy use. Gradual and steady improvements in the efficiencies of energy conversion and utilization have shown up as historical trends indicating decreasing ratios between per capita energy consumption and per capita real GNP. The life-style preferred by a given society will also enter into the equation. The level of comfort in terms of temperatures maintained during summer and winter plus the levels of illumination in residential, commercial, and industrial structures is but one example. The preference for suburban living, mobility, and decreased reliance on mass transit systems furnish another, and both examples, in their own way, can lead to increases in per capita energy consumption without a compensating increase in economic productivity. There are far too many other potential factors that could be listed here – the growing energy intensiveness of agriculture and the food industry, and the displacement of natural products by synthetics, are again only two examples.

Fig. 1.1 illustrates these "reservations" about the correlation between GNP and energy consumption. Here countries having nearly the same GNP per capita show wide variations in the per capita energy consumption (whether they are low-income, middle-income or industrialized countries). Furthermore, Fig. 1.1 shows that centrally planned economies with GNP per capita lower than that of the oil exporting countries have a higher per capita energy consumption. On the other hand Fig. 1.2 illustrates that energy consumption could be made independent of the GNP per capita, through more efficient and rational use of energy. Constraints placed on energy consumption in the 1970s in the U.S.A., for example, did not affect the growth in the per capita GNP; on the contrary the GNP per capita increased by about 1.7-fold between 1972 and 1978 while the per capita energy consumption remained nearly constant.

Fig. 1.1 Relationship between GNP/Capita (US$) and energy consumption (1978). Small dots represent individual countries. Large dots weighed averages (data after World Bank, 1980).

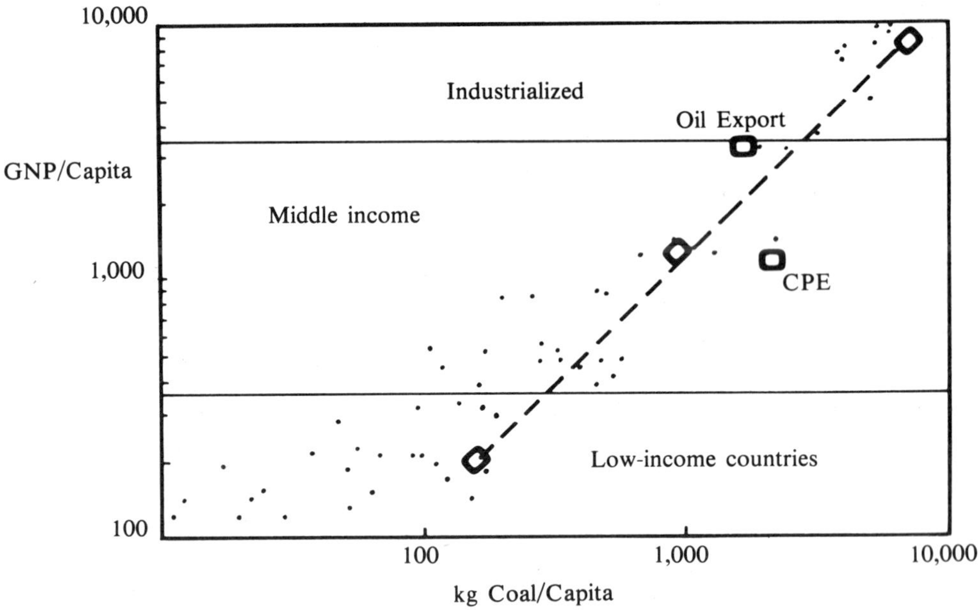

The increase in energy consumption in the past fifty years or so was accompanied by a major change in primary energy sources. In the 1920s coal accounted for about 80 per cent of the world's total commercial energy consumption; its share in later years was greatly reduced due to the increase in oil discoveries. Numerous technological changes took place to give impetus to this shift from coal to oil. A whole set of new technologies were developed not only to improve methods of oil supply but also to expand the range of its end uses. Advances in oil exploration and production technology improved the supply capacity of the oil industry; improvements in transport of oil by pipelines and tanker technology added to the advantage of oil as an easily transportable fuel; and the development of refining technology contributed greatly to improving efficiency in its end uses. All these factors worked to strengthen the competitiveness of oil as an alternative fuel to coal. As regards end uses, the widespread use of internal combustion engines, particularly in transport and in power generation, greatly increased the demand for oil. The result of this shift to oil was a drop of coal's share in the world's commercial energy consumption to 61% in 1960 and to 35% in 1970. The major transition to oil and natural gas took place in the early 1960's.

Fig. 1.2 Relationship between GNP/Capita (US$) and energy consumption in the U.S.A. (1972–1978) (data from World Bank, 1979, 1980 and UN 1980).

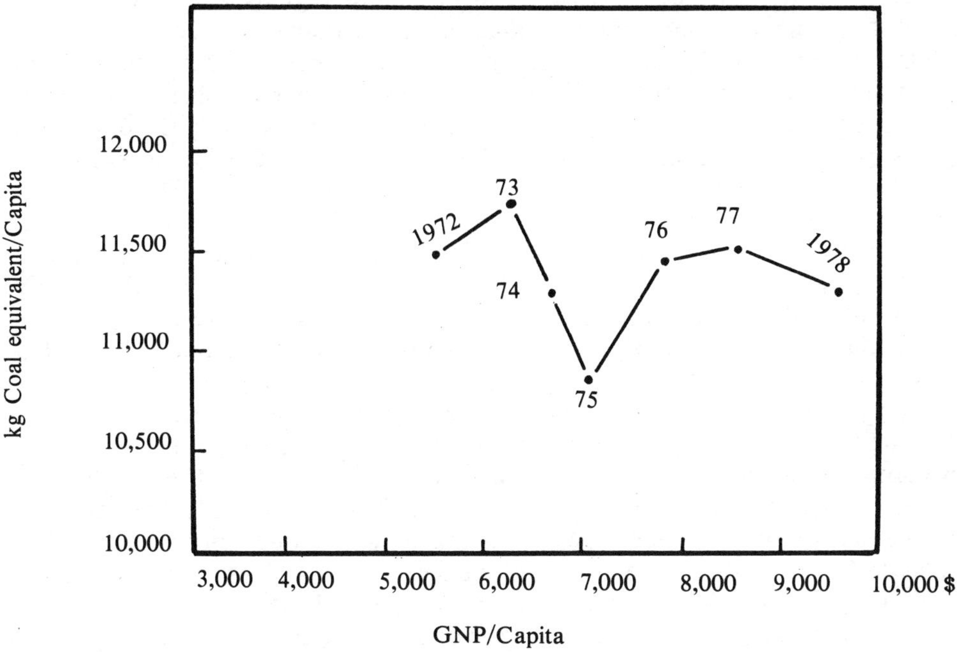

The consequences of the shift to cheap oil* were far-reaching. The shift implied not only the creation of physical and economic structures based on the use of oil but also the spread of particular social institutions and cultural values that made life dependent on its ready availability (see, for example, Pendse, 1979). Obviously it is in the developed countries, where most of the aforementioned technological changes and new patterns of energy consumption took place, that the shift from coal to oil and its consequences have been most visible. The considerable amounts of oil thus consumed in these countries is a clear reflection not only of their dependence on oil but also of a wasteful use of a low-priced resource. In developing countries, however, dependence on oil is even more complete, since these countries started their industrialization and development efforts after the shift from coal to oil had been well established. Not

* In 1950, the posted price of a barrel of oil in the Middle East was about $2.0; it dropped to $1.8 where it remained until 1970.

only were oil-dependent structures built up in these countries through imports of oil-geared capital goods from developed countries, but the latter's oil-based pattern of life was also taken over to a significant degree.

Disparity in Energy Consumption

Consumption of the world's commercial energy resources is heavily concentrated in the developed regions – the developed market economy countries and the socialist countries of Eastern Europe. These regions, with about 30 per cent of the total world population, account for more than 80 per cent of the total world consumption of commercial energy; the other 70 per cent of the population, comprising the developing countries and the socialist countries of Asia, consume less than 20 per cent. This heavy concentration in the developed regions is a reflection of the more advanced level of development already achieved. Their industries are characterized by energy-intensive processes; extensive coverage of communities and individual households by power grid systems provides ready access to energy for the daily life of people for heating, cooking, lighting and domestic work; and mechanization has reached not only farming and factory work but also commercial facilities. As a result, per capita commercial energy consumption in the developed countries is about 12 times that of the developing countries.

In the developing regions commercial energy consumption is relatively low, reflecting the low level of industrialization and general low economic development. Consumption of commercial energy in these countries is typically concentrated in urban areas, where the beneficiaries are mainly industry, the commercial sector and those few belonging to upper and upper-middle class societies. In rural areas and in urban slums, where the vast majority of the population live, most household energy needs are still met by non-commercial energy such as fuelwood, agricultural residues and animal manure, as well as human and animal power, rather than by commercial sources such as electricity and oil. Although no reliable national and/or global statistics exist on the current use of non-commercial sources of energy, estimates selected countries seem to indicate that these energy resources occupy an important share of the total energy consumption of developing countries. To mention a few recent estimates: non-commercial energy represents 90 per cent of all energy consumed in Nepal (Karki and Coburn, 1977), 83 per cent in Bangladesh, and 48 per cent in India (Hendersen, 1975). Unfortunately, non-commercial sources of energy have, hitherto, been largely ignored by many energy analysts. This has led to gross underestimations of the total energy consumption (or needs) of the developing countries. Energy consumption in India is a case in point. If only commercial energy is used as a measure, India's energy consumption is its recorded 1.7 million barrels a day of oil equivalent (MBDOE). However, if the use of two non-commercial energy sources – fuelwood and animal dung – is also considered, consumption rises to 4.1 MBDOE. Finally, if animal draft power (a significant source of energy in rural areas) is added, India's total energy consumption is actually 5.1 MBDOE, or triple its consumption of commercial energy alone (Howe and Tarrant, 1980).

Events of the 1970s

Although in the 1960s some experts voiced concern about the world's future energy problems, such warnings were not taken seriously until the events of the early 1970s (environmental awareness and anti-pollution campaigns, increases in oil prices, oil embargoes) brought home the reality of the problem. At that time the term "energy crisis" came into common use, and a great deal of confusion has been created about what "crisis" actually means, what constitutes the crisis, and whether there is one crisis or many.

The increases in oil prices since the early 1970s have had several far-reaching consequences, some positive and others negative. On one hand, cheap oil dissuaded many governments, particularly in developing countries, from planning their energy needs, from establishing conservation measures, and more important from accelerating research and development to harness their own energy resources (Attiga, 1979; Pendse, 1979). In India, for example, the marine geophysical surveys of the offshore oil field known as Bombay High were conducted, and the presence of oil clearly indicated, as early as 1963. But it took another seven years for detailed seismic surveys to confirm the availability of oil. Work began in earnest only after the oil price increase, and the oilfield is alreAdy yielding an output at the rate of 4 million tones per annum. In Malaysia too, offshore oil production received attention only after the oil price increase and output has doubled to reach more than 200,000 barrels per day. Even Britain's North Sea oil can be said to have passed through a similar process. Seismic prospecting was going on in the early 1960s; but oil production really started increasing from 1975, when it was 1.6 million tonnes, to 12 million tonnes in 1976 and to about 38 million tonnes in 1977 (Pendse, 1979). Efforts to exploit alternative sources of energy were also not worthwhile, and research and development were not vigorously pursued in the cheap oil era. However, since the increase in oil prices in the early 1970s extensive research and development programmes have been initiated by many countries to harness alternative sources of energy, particularly renewable ones. Furthermore, increasing efforts are being devoted to develop ways and means for more efficient production and consumption of energy.

On the other hand, the sudden increase in oil prices has created "confusion" in the world economy. The impact has been varied on developing countries, leading to serious fuel shortages in some areas and to curtailed industrial activities and even cutbacks in essential services in others, particularly in some rural areas. Higher oil prices have brought higher fuel and fertilizer costs for agriculture and higher costs for boiler fuel and gasoline for industrial production, electricity generation transportation. Sudden increases in the prices of kerosene – and reduced availability in some cases – have forced many people in West Africa and elsewhere back to charcoal and fuelwood for cooking fuel (which helps force up the prices of these commodities). The rise in energy prices has also meant higher prices for all goods imported from industrialized countries, thereby affecting international trade trends (Howe and Tarrant, 1980). Trehan *et al.* (1979) found that developing countries that depend on imported oil suffered severe repercussions: a drop by more than half in the real GNP per capita growth rate (from 3.1% to 1.5%) after 1973; a more than three-fold increase in inflation rates (from 7% to 24.5%) and a three-fold increase in real balance of payment deficits after 1973. However, most developing countries will continue to import much of their energy needs. Their bill for imported oil for energy use rose in nominal terms from $22 billion in 1975 to $29 billion in 1978, and is expected to increase to about $107 billion in 1985 and about $200 billion in 1990 (World Bank, 1980).

Locally, nationally and sometimes regionally, the environmental aspects of energy production and use were widely debated during the 1970s. Environmental awareness and anti-pollution campaigns affected the formulation of energy policies in many countries*, and nations realized that they are not alone environmentally: the actions of one country may affect the environment in a neighbouring one. It has also been recognized that environmental objectives are neither inconsistent with energy policy, nor impose constraints upon it: a balance can be maintained between the need to preserve and improve the quality of the environment and the socio-economic goals and needs that depend on the availability of energy. In traditional societies, for example, fuelwood is the principal source of energy. Given its consumption pattern, the effect of rapid population and economic growth is to intensify the demand for wood, widen the gap between demand and supply and introduce incentives in favour of more intensive wood cutting. Given the localized character of wood supply, intensive wood cutting exerts concentrated pressure on tree cover and other woody vegetation that leads to deforestation. The impact of this is not just a rapid increase in the prices of fuelwood and all forest-related products and services, but more broadly, an over-all decline in agricultural productivity through such phenomena as landslides, flooding, soil erosion and even desertification. A greater understanding of the causal links operative in this sequence would enable identification of points of leverage where policy could induce cost-effective substitutions towards environmentally more appropriate energy consumption and resource use patterns. The broader perspective would highlight not just price policies but also policies designed to promote better resource management and lower population density in areas with low carrying capacity. Moreover, it would facilitate the formulation of integrated community level programmes designed to meet localized needs. Clearly, to the extent that such policies and programmes succeed in preventing productivity decline, the dual objectives of economic growth and environmental protection are met.

* Gladwin *et al.* (1980) give a list of 803 cases of energy-environment conflict in OECD countries. One-fourth of the conflicts involve nuclear power plants; 16% oil refineries and petrochemical plants; 12% energy transportation and storage facilities and the rest other facilities. Sixty per cent of these conflicts were in the U.S.A., 12% in the UK, 5% in Canada, 4% in Japan, 4% in the Federal Republic of Germany and the rest in other OECD countries. See also El-Hinnawi (1980) for issues relating to nuclear power.

In summary, it could be pointed out that the 1970s brought into focus the following important issues: (1) a general realization of the finite nature of fossil fuels (oil and natural gas); (2) a general realization that the era of cheap energy was over and that all economies would have to adapt to high energy prices; (3) importance of developing indigenous energy resources and of establishing energy "mixes" to meet the demands for future development; (4) importance of the energy/environment relationship; (5) importance of increasing the efficiency of energy production and use (energy conservation), and (6) the realization that national development plans must be based on a conscious environmentally-sound energy policy. There is no doubt that in the 1970s, understanding of the nature and dimensions of the energy problem changed drastically for scientists, policy makers and the public. What were once considered exclusively technical problems amenable to technical solutions are now viewed correctly as facets of the larger problems that pervade all society. To state that energy policy questions are social, environmental and political as much as technical is now to state the obvious.

FUTURE ENERGY SUPPLY AND DEMAND

Until recently, national development plans were not based on a conscious energy policy but were geared to make various changes in the economy: rapid industrial development and capital-intensive GNP growth emphasizing short-term quantitative output rather than long-term resource utilization; increases in standards of living; cheap energy to improve competitive positions in an increasingly internationalized global economy; orientation of production towards energy-intensive industries like plastics and petrochemicals, and toward the mechanization of agriculture. The trend towards increasingly intensive energy use and thus increased oil consumption was an important stimulant to economic growth, which involved important changes in the structure of the economy and in the consumption habits and expectations of the public. The 1973–74 escalation of the price of crude oil marked the end of the era of inexpensive energy which primarily benefited those countries that had an early start in the Industrial Revolution. In future, rapid economic growth rates will be contingent to a substantial degree on the price and availability of energy resources, while in turn economic development and industrial modernization will modify the demand for energy, both nationally and globally.

While most governments have responded expeditiously to the energy "crisis", energy policy formulation and implementation is seriously handicapped by a wide range of the following uncertainties: (1) the magnitude of total energy demand as well as its profile, which results in estimates for the year 2000 diverging from the lowest to the highest figures by a factor of three and more for total demand and even more for certain fuels; (2) the quantity, location, and availability of various energy sources under a broad range of assumptions, including future energy prices; (3) the difficulty of assessing acurately the technical, environmental and economic aspects of various alternative energy systems within a time frame relevant to policymakers today; (4) interrelations between economic development planning and energy policy, including industrial location, rural electrification, urbanization and transition from non-commercial to commercial fuels; and (5) the geo-politics involved in energy supply and international trade.

Future Energy Demand

A distinction should first be made between "demand" and "need". Demand is that portion of energy need that becomes visible in commercial transactions; demand for energy depends on the customers having the means to buy it. Thus, energy demand in the developing countries is probably much smaller than actual energy needs, while demand in many industrialized countries is frequently considered to far exceed real needs (Howe and Tarrant, 1980).

Energy "need" is a subjective concept, and there is little agreement on how it should be defined. Several attempts have been made to date to quantify the "minimum" energy needs of people. Reddy and Prasad (1977) estimated the minimum per capita energy need to be 117×10^3 kJ/day, the level of achieving a "non-wasteful' standard of living similar to that in the U.S.A. Palmedo *et al.* (1978), on the basis of an indicator of well-being* introduced by Morris and Liser (1977), estimated the minimum energy needs to be in the range of 96 to 117×10^3 kJ/day per capita. Parikh (1978) estimated the energy essential for "subsistence" needs to be about 25×10^3 kJ/person/day. On the basis of the Bariloche Model (the Latin American World Model which explores possible physical limits to establish a society in which basic human needs are satisfied and, on the basis of a simple econometric model, investigates the possibility of doing so with current economic resources), Krugmann and Goldemberg (1980) estimated the basic energy needs in developing countries to be about 125×10^3 kJ/day/capita. It is doubtful that such numbers can be useful in planning or estimating future energy "needs" (Howe and Tarrant, 1980), because of the wide variations in assumptions, levels and patterns of development in different countries, etc. However, if we assume that the average basic energy needs is 120×10^3 kJ/day/capita, it is clear from the data given in Table 1.1 that such needs are not met in low-income and middle-income countries (developing countries). The data also show that energy consumption in industrialized countries is nearly five times such basic energy need.

* The indicator is a composite index, called Physical Quality of Life Index (PQLI) and ranges from 1 to 100. The Index is calculated for a given country by rating figures for life expectancy, infant mortality, and literacy each on a scale of 1 to 100 and averaging the three components, giving equal weight to each of them. Palmedo *et al.* plot the PQLI against per capita energy consumption, and suggest that the energy consumption range of 96 to 117×10^3 kJ/day per capita, where the marginal benefit of added energy drops sharply may be interpreted as representing a situation in which basic human needs are satisfied.

Table 1.1 Energy Consumption in 1978 (in 10^3 kJ/capita/day).

Low-income countries	13
Middle-income countries	72
Industrialized countries	561
Oil-exporting countries	129
Centrally-planned Economies	168

calculated from World Bank (1980) data.

Table 1.2 Estimates of Future World Primary Energy Demand (in million tonnes oil equivalent per year).

Source	1985	1990	2000	2020	2030
WAES (1977)(1)					
Scenario C	6160	—	—	—	—
D	5705	—	—	—	—
C_1	—	—	9895 (14500)	—	—
C_2	—	—	10340 (14950)	—	—
D_7	—	—	7995 (12570)	—	—
D_8	—	—	8270 (12857)	—	—
WEC (1978)(2)					
Scenario H–1	—	—	18182	41614	—
H–5	—	—	13182	22500	—
World Bank (1979)(3)	8945	11085	—	—	—
World Bank (1980)(4)	8300	10075	—	—	—
Global 2000 (1980)(5)					
High	8928	10666	—	—	—
Medium	8381 (7976)	9857 (9143)	—	—	—
Low	7833	9024	—	—	—
OECD–Inter-future (1980)(6)					
High	—	—	12002	—	—
Low	—	—	14622	—	—
IIASA (1980)(7)					
High	—	—	11801	—	24984
Low	—	—	9524	—	15691

NOTES:

(1) C: High economic growth, 6% per year; oil price $11.5/bbl, vigorous national policy; D: Low economic growth, 3% per year; oil price $11.5/bbl, restrained policy; C_1: High economic growth, 5% per year; oil price 17.25/bbl; C_2 with a limited OPEC production; D_7 emphasizes coal as replacement fuel while D_8 emphasizes nuclear energy.

WAES (1977) gives estimates for WOCA (World outside Communist Areas). Using additional data for Centrally Planned Economies, as given in Global 2000 (1980), the estimates for the world as a whole are shown in the above table between brackets.

(2) World Energy Conference (WEC, 1978) scenario H–1 and H–5 indicate high economic growth, 4.2% per year: H–1, no price response; H–5 high price response plus energy constraints.

(3) Assuming that the average annual economic growth rate 1980–1990 will be 5.6% for developing countries, 4.2% for industrialized countries, 5.0% for oil-exporting countries, 4.2% for Centrally Planned Economies, and that oil prices would remain at their average level for the period 1975–78.

(4) Assuming that the average annual economic growth rate 1985–1990 will be 5.6% for developing countries, 4.0% for industrialized countries, 5.8% for oil-exporting countries, 4.6% for Centrally Planned Economies.

(5) Assuming a constant oil price of $13/bbl in 1975 dollars through the year 2000. The medium case assumes medium population and GNP growth; the high case assumes low population and high GNP growth; and the low case assumes high population and GNP growth. Assuming that oil prices will rise at a real rate of 5% per year between 1980 and 1990, the estimates of energy demand are those given between brackets for the medium case.

(6) After Malkin (1980).

(7) Assuming different GDP per capita for different regions and different population projections (see IIASA, 1980).

Several studies have been made to estimate future energy demand in the world (Table 1.2). It should be noted that these and other energy "forecasts" rest on a multiplicity of different assumptions and different aggregating procedures that make them *roughly indicative* and subject to a wide range of error and changes. The main elements affecting these forecasts are (*a*) world and regional scenarios of economic development; (*b*) correlation of economic growth with energy consumption; (*c*) physical, economic, environmental and geo-political constraints applying to energy production and consumption; (*d*) future prices of different sources of energy; (*e*) future availability of different sources of energy and the development of appropriate technologies to harness them; (*f*) public acceptance of energy source and conservation measures. In addition, an inherent deficiency in all these "forecasts" is that they neglect non-commercial sources of energy which constitute a major resource in developing countries. Even by concentrating only on commercial sources of energy a wide range of error is encountered in assumptions pertaining to developing countries, especially in relation to the rate of substitution of non-commercial sources of energy by commercial ones.

THE POTENTIAL OF ENERGY RESOURCES

Coal

The total resources of solid fossil fuels (different types of coal and peat) are estimated at 11,184 × 10^9 tce (tonnes coal equivalent). Of these about 1,081 ×

10^9 tce are proved reserves (of which 693.2 × 10^9 tce are recoverable) and the rest is additional resources in situ*. Table 1.3 gives the estimated world resource of solid fossil fuels.

Table 1.3 World Resources of Solid Fossil Fuels (Coal and Peat).

Regions	Proved recoverable reserves		Additional resources in place	
	10^9 tonnes coal equiv.	%	10^9 tonnes coal equiv.	%
Africa	32.6	4.7	146.5	1.5
America	200.2	28.9	2,970.0	29.4
Asia	116.1	16.7	1,454.1	14.4
USSR	169.1	24.4	4,469.8	44.2
Europe	138.8	20.0	448.2	4.4
Oceania/ Australia	34.4	5.3	613.8	6.1
Total	691.2	100.0	10,102.4	100.0

After World Energy Conference (1980).

At the present rate of coal consumption (about 3,000 million tce), the proved recoverable reserves would last for about 230 years, while recoverable resources (total resources at 50% recoverability) would last approximately 1,800 years. The present proved recoverable reserves of 693 billion tce will meet the production planned for the year 2000, but not for the year 2020, if marked expansion in coal utilization is undertaken in the future. According to WEC (1978), if the production of 8,800 million tce by the year 2020 is to be achieved, the proved recoverable reserves have to be of the order of 1,200 billion tce. Since the potential of the coal reserves is so huge, it is very likely that the figure of the present proved recoverable reserves (693 billion tce) will double by the year 2020.

Oil

In the last 10 years the majority of estimates made for the total quantity of oil recoverable (ultimate recovery) have been in the range 240–360 × 10^9 tonnes (Table 1.4). Up to the end of 1978, aproximately 53 × 10^9 tonnes (15%) of the total recoverable oil (354 × 10^9 tonnes according to WEC, 1980) had been

* Proved reserves represent the fraction of the total resources that has been assessed as being exploitable with existing technology. Recoverable reserves are the fraction of proved reserves that can be recovered under existing economic and technological limits. Additional resources are those of at least foreseeable economic interest.

recovered. On 1.1.1979 the estimates for the proved recoverable reserves and the additional resources were 89 × 10^9 tonnes and 212 × 10^9 tonnes respectively (Table 1.5).

Table 1.4 Estimates of Total Recoverable Oil.

	10^9 tonnes
1970 Moody (Mobil)	246
1971 Warman (BP)	164–273 (218)
1971 Weeks	312
1971 US National Petroleum Council	364
1972 Linden	402
1972 Weeks	498
1972 Moody, Emerick (Modil)	246–259 (252.5)
1973 WEC (USGS)	184–1840
1975 Adams and Kirby (BP)	273
1975 Moody (Mobil)	277
1975 BGR	336
1976 Klemme (Weeks)	259
1978 WEC	350
1979 Halbouty (Moody)	304
1979 Meyerhoff	300
1979 Roorda	330
1980 WEC	354

After WEC (1980).

Table 1.5 Cumulative Production, Reserves, Resources, and Ultimate Recovery of Oil in Different Regions.

Region	Cumul. Production 1.1.1979		Proved recoverable reserves 1.1.1979		Estimated add recoverable resources		Ultimate recovery	
	million tonne	%	million tonne	%	million tonne	%	million tonne	%
Africa	3,750	7	8,040	9	34,000	16	45,790	13
North America	17,520	33	4,480	5	24,000	11	46,000	13
Latin America	7,040	14	7,770	9	12,000	6	26,810	8
Far East/Pacific	1,720	3	2,390	3	12,000	6	16,110	4
Middle East	14,680	28	51,050	57	52,000	24	117,730	33
Western Europe	560	1	2,710	3	10,000	5	13,270	4
USSR, China, Eastern Europe	7,530	14	12,700	14	64,000	30	84,230	24
Antarctic					4,000	2	4,000	1
Total	52,800	100	89,140	100	212,000	100	353,940	100

After WEC (1980). On 1.1.1980, the total proved recoverable reserves of oil were 87,894 million tonnes; on 1.1.1981, the total was 88,839 million tonnes (as calculated from Oil and Gas J., 1979 and 1980 respectively).

At the rate of oil consumption (about 3,000 million tonnes) the total world proved recoverable reserves (known on 1.1.1981)* would last for only 30 years; i.e. until the year 2010, and the total additional recoverable resource about 70 years.**

Natural Gas

Estimates made in the 1970s of ultimate recovery of natural gas vary between 200–300 × 10^{12} m^3; a recent estimate by WEC (1980) is 293 × 10^{12} m^3. Of this quantity a total of about 27 × 10^{12} m^3 (9%) had already been recovered by the end of 1978. Estimates for the proved recoverable reserves at the end of 1978 were about 74 × 10^{12} m^3 and for the additional recoverable resources 192 × 10^{12} m^3; see Table 1.6.

Table 1.6 World Natural Gas Resources (in 10^{12} m^3).

Region	Cumulative production up to 1.1.79	Proved rec. reserves on 1.1.79	Estimated additional Rec. Recources	"ultimate recovery"
Africa	0.1	7.3	26	33.4
North America	16.9	7.5	42	66.4
Latin America	1.8	4.7	10	16.5
Far East/Pacific	0.2	3.3	10	13.5
Middle East	1.1	20.5	30	51.6
Western Europe	1.5	3.9	6	11.4
USSR, China, Eastern Europe	5.2	26.9	64	96.1
Antarctic			4	4.0
Total	26.8	74.1	192	292.9

After WEC (1980). On 1.1.1980 the estimated proved recoverable reserves were 72 × 10^{12} m^3. While on 1.1.1981, they were 73.9 × 10^{12} m^3 (calculated after Oil and Gas J., 1979 and 1980, respectively).

At the present consumption rate (about 1.5 × 10^{12} m^3), the proved recoverable reserves of natural gas would last for about 49 years and the estimated additional resources for about 130 years.

Oil Shale and Tar Sands:

Some studies have been caried out to estimate the amounts of exploitable oil shales and tar sands in different countries (WEC, 1980). Estimates of the

* Does not include about 1,500 million tonnes natural gas liquids.
** Does not include 11,600 million tonnes natural gas liquids estimated by WEC (1980).

recoverable oil from these deposits amount to 46,262 million tonne and 40,051 million tonne, respectively.

Nuclear Energy

There has been a number of estimates of future nuclear energy production. However, such estimates are subject to large margins of uncertainty and can provide no more than very general indications (El-Hinnawi, 1980, 1981). The IAEA estimates that the total nuclear power growth in the world will be in the range of 1,100–1,700 GW(e) by the year 2000. Haefele (1979) gives an absolute upper limit of 10,000 GW(e) by the year 2030.

The nuclear power industry's need for uranium in the short term are considerable, and indicate the need for accelerated efforts to expand current resources. The requirement for uranium in 1977 was slightly less than 30,000 tonnes (U) and is expected to reach 178,000 tonnes (U) in the year 2000 (OECD, 1978). The cumulative requirements of natural uranium by the year 2000 for the no-recycle scenario (2,276,000 tonne U) are within known world uranium resources, currently estimated to total about 4 million tonnes in the cost category less than $130/kg U. However, considerable exploration effort will be required before such quantities of uranium can be classified as "reserves", the resource category normally considered for the purpose of production. The cumulative uranium requirements by the year 2025 will reach 9 million tonnes U for light water reactors without recycle (OECD, 1978), which greatly exceeds the currently estimated world resources of 4 million tonnes. At present, there is no concensus as to whether low-cost uranium to this amount actually physically exists in the uppermost part of the earth's crust from which it could be economically produced. Taking the world proven reserves of uranium as 4 million* tonnes, the resource base for nuclear energy is abour 60×10^{12} W (equivalent to 42×10^{9} tonne oil equivalent). The development of alternative nuclear fuel cycles, including reprocessing and the deployment of the breeder reactor (El-Hinnawi, 1980, 1981), could extend the resource base by about 30%. However, it is not evident that such developments will occur in the near future, at least not on a large scale within the next two to three decades.

Renewable Sources of Energy

The resource base of renewable sources of energy is extremely large with an infinite potential for renewal. With the present state of technology, it is, however difficult to estimate how much of the resource base can be technically and economically exploited, and what would be the contribution of renewable sources of energy to the overall future world energy supply systems.

White (1965) estimated the heat stored, beyond surface temperatures, in the outer 10 km of the earth's crust to be about 12.6×10^{26} J. This resource base is

* Haefele (1979) gives a figure of 20 million tonnes as the global proven regional uranium resource densities. This is a grossly overestimated figure that cannot be practically exploited.

equivalent to the heat content of 4.6×10^{16} tonnes of coal, or more than about 70,000 times the heat content of the technically and economically recoverable coal reserves in the world. However, such geothermal heat is too diffuse to be exploited on a worldwide basis. Resources suitable for commercial exploitation are confined to suitable geological deposits mostly associated with tectonic activity. It has been estimated that by the year 2000 about 76 to 100 GW(e) and about 76 GW(t) of geothermal energy could be harnessed worldwide (WEC, 1978).

The annual amount of solar radiation received at the surface of the earth is about 1.2×10^{17} W (or about 1×10^{18} kWh). The extraction of all the energy provided by the sun is, however, nearly impossible – only a fraction can be extracted: the efficiency of extraction depends on the location and the prevailing meteorological conditions.

Different authors have estimated the global wind energy resource base to be in the range of $1{,}200 \times 10^{12}$ W (Sørensen, 1980), but the harnessing of this power is very site-dependent. Sørensen (1979) estimated the wind potential along continental coastlines (Table 1.7) and Merriam (1978) estimated that by the year 2000 wind could contribute 9×10^{11} kWh of electricity per year.

Table 1.7 Estimate of Coastline Wind Potential.

Region	Assumed Coastline (million km)	Est. Coastline Wind Potential (TWh per year)	Present electricity Consumption (1978) (TWh per year)
North America	74	754	2400
South America	35	604	300
Oceania	32	780	150
USSR	18	494	1000
Asia (excluding USSR)	67	701	1000
Europe (excluding USSR)	38	1051	2000
Africa	43	534	200

After Sørensen, 1979; consumption figures have been rounded. TWh = terawatt hours. 1 TW = 1 trillion (10^{12}) watts.

The energy and power in waves can be considerable. A typical mid-ocean has approximately 1.5 m waves at periods of 8 seconds (see Chapter 5). These waves correspond to a mean flux of wave power across a section of the ocean of the order of 10 kW/m. Summed over the entire ice-free ocean, the total power available would be 2.7×10^{12} W; the amount of power that could be extracted would, however, be a small fraction of this amount. Similarly, the power available at tidal sites is about 60×10^{9} W, but no more than 10–25% can be converted into electricity.

It has been estimated that by placing Ocean Thermal Energy Conversion facilities approximately 15 km apart throughout the oceans between 20°S and 20°N latitude, the theoretical maximum limit for electricity production is of the order of 50×10^{12} W. An ultimate installed capacity representing a small fraction of this potential would still represent a significant contribution, especially to the energy needs of islands and isolated coastal regions (see Chapter 5).

Armstrong (1978) estimated the total potential from hydro resources of the world at 2.2 million MW of installed and installable generating capacity of 50% capacity factor. These estimates are, however, conservative and do not take fully into account the great potential of small hydropower schemes. The operating hydroelectric capacity at present is about 400 GW with an annual production of about $1,558 \times 10^9$ kWh, which is approximately 35% of the total world potential available 95% of the time (estimated to be $4,434 \times 10^9$ kWh/y; see Chapter 6).

The dry weight of all living biomass (plant matter) on the earth's land surface has been estimated as nearly $2,400 \times 10^9$ dry tonnes, with a primary production rate of 171×10^9 dry tonnes/y (Rodin *et al.*, 1975). The latter is equivalent to more than 10 times the present total world energy consumption from all commercial sources (assuming that the average calorific value of dry biomass is about 1,900 kJ/kg). However, only a modest fraction of the biomass is used for fuel (at present mostly fuelwood, agricultural residues, cow-dung, etc.). The fuelwood consumption in 1978 is estimated to be about 1,720 million tonne; no reliable estimates are available of the consumption of other biomass for fuel.

FUTURE ENERGY SCENARIOS

As mentioned earlier, estimates of future energy demand are subject to a wide range of uncertainty because of the different and partly unpredictable assumptions on which these estimates are based. The same is true for the future world energy supply. Future production is influenced by new discoveries, improved production techniques, development of new energy sources, economic, geo-political and environmental constraints, and other factors.

Different scenarios have been suggested to balance future energy demand and supply (e.g. WEC, 1978; WAES, 1977; IIASA, 1980), but the question of how to meet future energy demand by an environmentally-sound and appropriate energy "mix" still remains unanswered. Such an energy "mix" will vary from one country to another depending on availability of resources, technology, level of development, economics, environmental considerations, public acceptability, etc. Both the WAES (1977) and WEC (1978) studies indicate that the supply of oil will fail to meet the increasing demand for oil before the year 2000. IIASA (1980) came to nearly the same conclusion but indicated that unconventional oil (from oil shales; tar sands; oil produced by enhanced recovery, from deep offshore well, etc.) will stretch the supply to meet the demand by the year 2020 and then the "gap" has to be filled by synthetic liquid fuels produced from coal.

The future shortage in oil supplies has implications for the whole world. But it is the developing countries with less financial resources that will lose out in the worldwide scramble for energy resources as the supply constraint increasingly becomes a reality. This is particularly alarming since the domestic energy requirements of developing countries are growing, and are expected to grow in the future, with increasing pace of industrialization and development (Table 1.8). How to ensure energy supplies for development, therefore, becomes in the long run a question of critical importance for the developing countries themselves and for the world as a whole.

Table 1.8 Energy Consumption in the Year 2000.

	Population (10^9)		Energy consumption (10^{15} kJ)		Energy Consumption increase by
	1975	2000	1975	2000	
World	3.94	6.21	263	890	3.5
Industrialized countries	1.12	1.36	202 (77%)	530 (60%)	2.6
Developing countries	2.82	4.85	61 (23%)	360 (40%)	6.0

After Goldemberg (1979).

Needless to say, the problem does not apply uniformly to all developing countries. In developing countries endowed with energy resources, the main concern may be how to optimize the use of their resources for development. Depending on the rate of depletion of their energy resource endowment, these countries may also have to seek alternative resources and adjust the end-use pattern energy accordingly. For developing countries already heavily dependent on imports of energy, which constitute the majority, the ensuring of an adequate supply will most certainly be a serious concern. These countries will also be hard pressed to chart their course out of their present dependence on oil.

In view of the difficulties to be encountered to meet future world energy demand, technological developments will be increaseingly necessary to maintain sufficient energy supplies. Alternatives include both new technologies and existing technologies that become economic as energy prices rise. Improvements could occur not only in actual exploration and production techniques but also in electric utility operations, direct fuel consumption, energy conversion, and efficiency levels. Determining the impact of various methods is complicated because each may penetrate the market at a different time, with varying implementation and success rates. Among the technologies expected to be competitive before the year 2000 are heat pumps, enhanced recovery of oil, and the development of some new and renewable sources of energy.

In this respect, it should be noted that some renewable sources of energy (for example, fuelwood, agricultural residues, cow-dung cakes, etc.) have been used, especially in rural areas of the developing countries, since millennia, but in a primitive and most inefficient way. Recently, attention has been given to develop technologies for the better harnessing and more efficient use of these and other renewable sources of energy. A great deal has been said about the potential and role of renewable sources of energy in meeting part of the future energy demand both in developed and in developing countries (see, for example; Hayes, 1973; Lovins, 1976; Weingart, 1979; El-Hinnawi, 1977, 1981). In the developed countries strategies for the exploitation of renewable sources of energy constitute a part of recent policies which aim at reducing the dependence on fossil fuels. Solar technologies (especially solar heating for buildings and for some industrial process heating), geothermal, technologies, production of fuel from biomass, etc. are becoming more and more common in several developed countries. A particularly important social advantage of the development of renewable sources of energy is the potential for promoting development in rural areas of the developing countries. It is here that hundreds of millions of people live far from the mainstream of economic activities associated with the process of development, a process that has, hitherto, unjustifiably been confined to urban and semi-urban areas. Faced with the harsh realities of poverty, disease, and starvation, the rural population embarked on an exodus to urban areas seeking better opportunities. Not only has this exacerbated the already dire urban problems but it has also delayed rural development and in some cases caused a reduction in agricultural productivity (through lack of man power and increased negligence of land). Such exodus can only be brought into a halt (and may even be reversed) by accelerated rural development programmes. At the heart of such programmes lies the provision of an adequate supply of energy.

Most of the energy consumed in rural areas of the developing countries is for cooking of food. The overwhelming share of this energy is non-commercial. The consequences of heavy reliance on non-commercial fuels are expanding deforestation, devegetation, declining water tables, soil erosion, silting, and flooding as forests are destroyed – and rising prices for wood and charcoal. Where wood or charcoal are marketed, the poor may spend as much as 30 per cent of a household budget for cooking fuel. And even such non-commercial sources of energy are not used in the most efficient way. Present cooking stoves and charcoal kilns convert wood and other organic wastes to energy very inefficiently. Furthermore, the inefficiently burned non-commercial fuels have valuable alternative uses. Crop residues and cow-dung provide vital organic materials and nutrients to the soil which cannot wholly be replaced by any quantity of expensive chemical fertilizers. Some crop residues are essential for feeding livestock, and trees and forests protect watersheds and modify the local climate.

Petroleum-based fuels and electricity are less widely available in rural areas and generally more expensive than in urban areas. In many cases, an "electrified" village is one in which only a few buildings (for example, the clinic, the school, a few homes of the village elite) have electricity. The gasoline and diesel fuel situation is similarly bleak. Few people in rural areas can afford to buy such fuel and the machinery (pumps, tractors, etc.) with which to use it. The majority cannot afford to do so and this contributes to the vicious circle of poverty: the poor who cannot afford energy or energy equipment cannot increase their productivity, and they remain poor. Yet through the use of appropriate technology and by proper management, a wide variety of renewable sources of energy can be harnessed to meet the essential needs of the poor offering them a better quality of life.

Unfortunately, although there is a growing awareness of the broad importance of energy questions to economic development, the immediate and overwhelming concern of policy-makers is not the question of long-term needs and supplies, but the question of how to pay for this month's oil imports or how to finance the construction of a large power station. Howe and Tarrant (1980) have rightly described the situation as follows "unfortunately, opinion among analysts as to the merits of small-scale renewable energy varies widely. Experts on these energy sources tend to become advocates – or even evangelists – as do experts on conventional large-scale energy. Those who support small-scale renewable energy are indignant at the risks being proposed by nuclear advocates, and the latter are shocked at the suggestions that solar technology be adopted in its present immature state. Each side extols the virtues of its own solutions and assails the claims of the other. There is too little of the spirit of honest inquiry". Policy-makers tend to believe that "large" is beautiful. This is especially true in developing countries where priority is given to construction of large power plants, which serve mainly urban and industrial centres, increasing thereby the disparity in energy consumption between people living in such areas and those living in rural areas. The policy-makers tend to forget that such "large" technologies penetrated the market only after years of extensive and expensive research and development. Weinberg (1979) pointed out that "Time and information, net energy, and capital cost impose constraints on the rate of introduction of new energy technologies that cannot be ignored by policy-makers. This is hardly new, nor startling. Indeed, the one energy technology that has penetrated despite these constraints, nuclear energy, did so because it was subsidized – by the government in the form of support for R & D and demonstrations, by reactor vendors who in aggregate in the U.S.A. have lost perhaps $\$10^9$ on the reactor business. The other new technologies will also require subsidy".

The evaluation of the potential development and use of renewable sources of energy should go much further than conventional financial cost-benefit analysis to include a comprehensive appraisal of social costs and benefits, despite the methodological problems involved in providing quantified assessments. Account must be taken of the needs of those who will be the end-users of these technologies. This means the public at large; it also requires assessment of the future energy needs of the next generation. Other factors such as cultural and social acceptance of change, and the relevant political, economic and social structures and processes should be considered through a comprehensive systems approach to the supply and use of energy in all forms. However, it needs to be emphasised that there are no absolute barriers to the greater use now of new and renewable sources of energy and any such barriers if they exist should be tackled effectively through appropriate and urgent national, regional and international action (see also, United Nations Conference on New and Renewable Sources of Energy, 1981).

References

Armstrong, E. L. (1978): Hydraulic resources. In World energy resources 1985–2020 World Energy Conference IPC Science and Technology Press, Guildford, U.K.

Attiga, A. A. (1979): Global energy transition and the third world. Third World Quarterly 1,4,39.

CEQ (1980): The global 2000 report to the President of the USA. Pergamon Press, Oxford.

El-Hinnawi, E. (1977): Energy, Environment and Development. 10th World Energy Conf. Istanbul, paper 4.8–1.

El-Hinnawi, E. (1980): Nuclear energy and the environment. Pergamon Press, Oxford.

El-Hinnawi, E. (1981): Environmental Impacts of Production and Use of Energy. Tycooly International Publishing Ltd., Dublin.

Gladwen *et al.* (1980): Energy/environment conflict. Paper prepared for OECD. OECD, Paris.

Goldemberg, J. (1979): Global Options for Short-Range Alternative Energy Strategies. Energy, 4. 733.

Haefele, W. (1979): Global Perspectives and Options for Long-Range Energy Strategies. Energy 4, 745.

Hayes, D. (1973): Rays of Hope. Norton, New York.

Henderson, P. D. (1975): India: the energy sector. Oxford University Press, New Delhi, India.

Howe, J. W. and J. J. Tarrant (1980): An alternative road to the post-petroleum era: north-south co-operation. Overseas Development Council, Monograph 14, Washington DC.

IIASA (1980): The comparison of energy options: a methodological study. Report by the energy systems Progress Group, IIASA, Laxenburg.

Karki, A. B. and B. A. Coburn (1977): The prospect of biogas as one of the sources of energy in Nepal. 10th World Energy Conference, Istanbul, Sect. 4.6–3.

Krugmann, H. and J. Goldemberg (1980): The Energy Cost of Satisfying Basic Human Needs. Instituto de Fisica, Sáo Paulo University, Brazil.

Lovins, A. (1976): Energy Strategy; the Road Not Taken. Foreign Affairs 55,1.

Merriam, M. F. (1980): Wind, waves and tides. Annual review of energy, Vol. 3, p. 29. Annual Reviews Inc., Palo Alto, California.

Morris, M. D. and F. B. Liser (1977): The PQLI: Measuring Progress in Meeting Human Needs. Communique No. 32 Overseas Development Council, Washington, DC.

OECD (1978): Nuclear fuel cycle requirements, OECD, Paris.

Palmedo, P. F. et al. (1978): Energy Needs and Resources in Developing Countries. Brookhaven National Laboratory, Report, BNL 50784.

Parikh, J. (1978): Energy and development. Public Utilities Report No. PUN43, World Bank, Washington, DC., p. 227–233.

Pendse, D. R. (1979): The energy crisis and third world options. Third World Quarterly 1,4,69.

Reddy, A. K. N. and K. K. Prasad, (1977): Technological alternatives and the Indian energy crisis. Economic and Political Weekly (India), special No. August 1977, p. 1496.

Rodin, L. E. *et al.* (1975): Productivity of the world's main ecosystems. Proc. 5th General Assembly, Special Commitee IBP. National Academy of Sciences, Washington, DC.

Sørensen, B. (1979): Wind energy. UNITAR Conf. on Long-Term Energy Sources, UNITAR, UN, New York.

Sørensen, B. (1980): Renewable energy. Academic Press, London.

Trehan, R. K. *et al.* (1979): Energy and development, a preliminary analysis for less developed countries. MTR–79Woo216. The MITRE Corporation.

United Nations (1980): 1979 Yearbook of World Energy Statistics. Sales No. E/F.80.XVII.7, United Nations, New York.

United Nations Conference on New and Renewable Sources of Energy (1981): Report of the Synthesis Group. A/Conf.100/PC/41. United Nations, New York.

WAES (1977): Energy: global prospects 1985–2000. Report of the Workshop on Alternative Energy Strategies. McGraw-Hill, New York.

WEC (1978): World energy resources 1985–2020. IPC Science and Technology Press, Guildford, UK.

WEC (1980): Survey of energy resources. World Energy Conference, Munich 1980.

Weinberg, A. M. (1979): Are the Alternative Energy Strategies Achievable? Energy, 4, 941.

Weingart, J. M. (1979): Global Aspects of Sunlight as a Major Energy Source. Energy, 4, 775.

White, D. E. (1965): Geothermal energy. US Geological Survey Circular 519.

World Bank (1980): World Development report. World Bank, Washington, DC.

World Bank (1979): World Development report 1979. Washington, DC.

CHAPTER 2

Geothermal Energy

A. J. ELLIS

Asistant Director-General
Department of Scientific and Industrial Research
Wellington, New Zealand

Fig. 2 The Wairakei Geothermal Field with power station in the foreground.

INTRODUCTION

GEOTHERMAL energy is based on the natural heat of the earth. The upper crust has a mean temperature gradient of 20–30° C/km depth; White (1965) estimated the heat stored, beyond surface temperatures, in the outer 10 km of the earth's crust, to be about $12 \cdot 6 \times 10^{26}$J. This resource base is equivalent to the heat content of $4 \cdot 6 \times 10^{16}$ tonne of coal (assuming the heat content of coal to be $27 \cdot 6 \times 10^{9}$J/tonne), or more than about 70,000 times the heat content of the technically and economically recoverable coal reserves in the world*. However, geothermal heat in the outer 10 km of the earth's crust is too diffuse to be an exploitable energy resource on a worldwide basis. Resources suitable for commercial exploitation may be defined as localized geological deposits of heat concentrated at attainable depths, in confined volumes, and at temperatures sufficient for electric or thermal energy utilization.

TYPES OF GEOTHERMAL RESOURCES

From the geological point of view, geothermal resources can be classified into: hydrothermal convection systems, hot igneous systems and conduction-dominated systems. Fig. 2.1 is a map illustrating the geographical zones with potential geothermal activity, which coincide normally with tectonic zones.

Subsurface reservoirs of steam or hot water, which may display such surface characteristics as boiling springs, sulphurous mud flats and fumaroles, are categorized as *hydrothermal convection systems*. The creation of such systems begins with a source of heat, hot or molten rock, that lies relatively close to the earth's surface. Overlying this high temperature rock zone is a permeable rock formation containing water, largely of meteoric origin, which rises upward as it is heated by the hot rock below (Fig. 2.2). Above the permeable rock is a cap of impermeable or semi-impermeable rock, which traps the super-heated water, but cracks or fissures in it allow the fluid to rise to the earth's surface either as steam (a vapour-dominated or dry steam system) or hot water (a liquid-dominated system). Hydrothermal convection systems are usually associated with the earth crustal tectonic plate boundaries and related volcanic activity.

Hydrothermal convection systems can be subdivided into hot water-dominated or vapour-dominated systems, depending on whether the near-surface permeable zones produce principally water or steam when tapped by production

*The coal reserves recoverable under current economic and technological conditions are estimated to be $6 \cdot 3 \times 10^{11}$ tonne (Peters et al., 1978).

Fig. 2.1 World Geothermal Regions

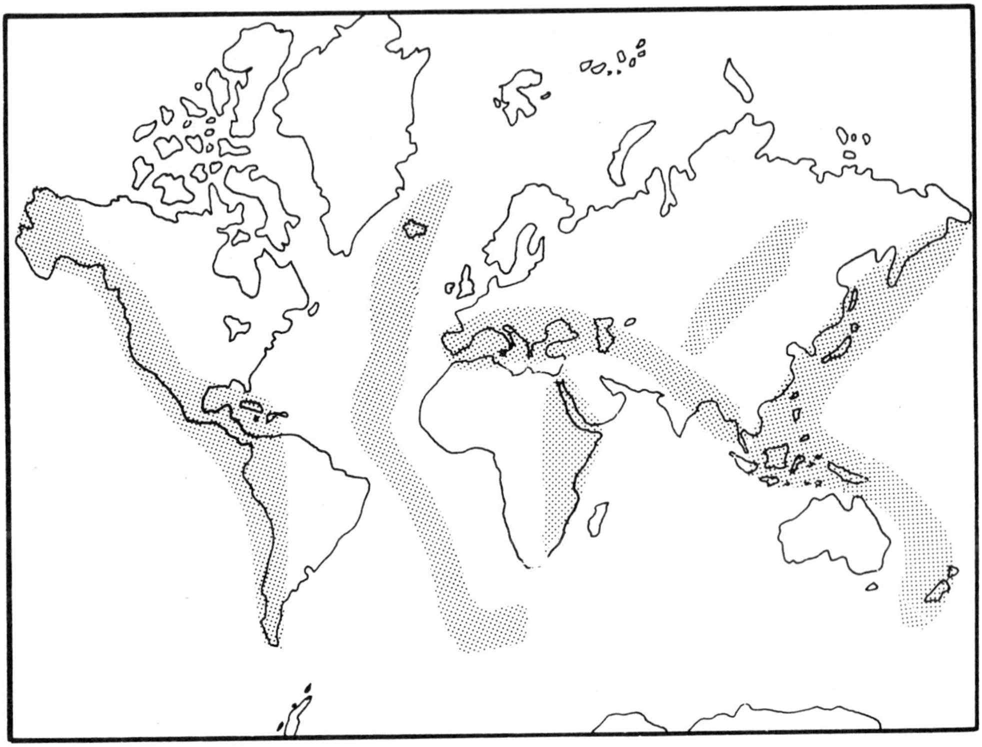

wells. Liquid-dominated systems (or hot-water systems), which are far more common than vapour-dominated ones, vary greatly in their chemical and thermal characteristics by site. Based on the temperature of the liquid, these systems can be classified into :

(*a*) High-temperature; greater than 150°C

(*b*) Medium-temperature; 90°C — 150°C

(*c*) Low-temperature; less than 90°C.

High-temperature systems may be subdivided according to their characteristics, namely, salinity, chemical composition, stratigraphic set-up, permeability of reservoir, etc. These characteristics are important in determining the commercial viability of a hot-water system, either for electricity generation or for other purposes.

Generally, the flashed steam process is used for electricity generation from hot-water systems. In this process (Fig. 2.3) as the hot water (which is under high pressure) is withdrawn from the reservoir by wells and nears the surface, the pressure decreases, causing about 20 per cent of the fluid to boil and "flash" into steam. The steam is separated from the water and then directed to the turbine.

The hot mineralized water leaving the separator may be further used or processed before disposal, preferably by reinjection into the local rock formation. It may be desalinated before further use or prior to reinjection (minerals may be economically recoverable). Examples of hot-water geothermal fields are Wairakei and Broadlands in New Zealand, Cerro Prieto in Mexico, Salton Sea in California and Otake in Japan.

Fig. 2.2 Schematic Diagram of a Geothermal Reservoir

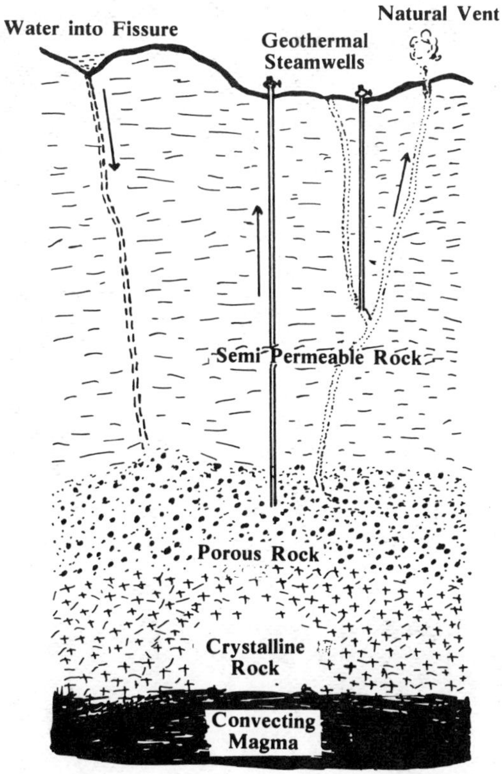

Another method for the production of electricity from high- or medium-temperature geothermal waters is the binary cycle process (Fig. 2.4). In this process, the water withdrawn from the reservoir is used to heat a secondary fluid (freon or isobutane) in a heat exchanger. The vapour thus generated by boiling the secondary fluid is used to drive the turbine. Once used, the vapour is condensed and recirculated through the heat exchanger in a closed system, where it may be heated and used again. A binary-cycle geothermal plant rated at 11MW(e) has been constructed on East Mesa, East of the Imperial Valley, California. Such plants can be used for production of electricity at suitable sites, especially in remote rural areas (Phéline, 1979).

Fig. 2.3 A Schematic Flow Chart for Power Production from a Hot Water Geothermal Field showing the Final Distribution of Soluble and Gaseous Chemical Constituents as Fluids Reach the Environment, or are Reinjected (after Ellis, 1975).

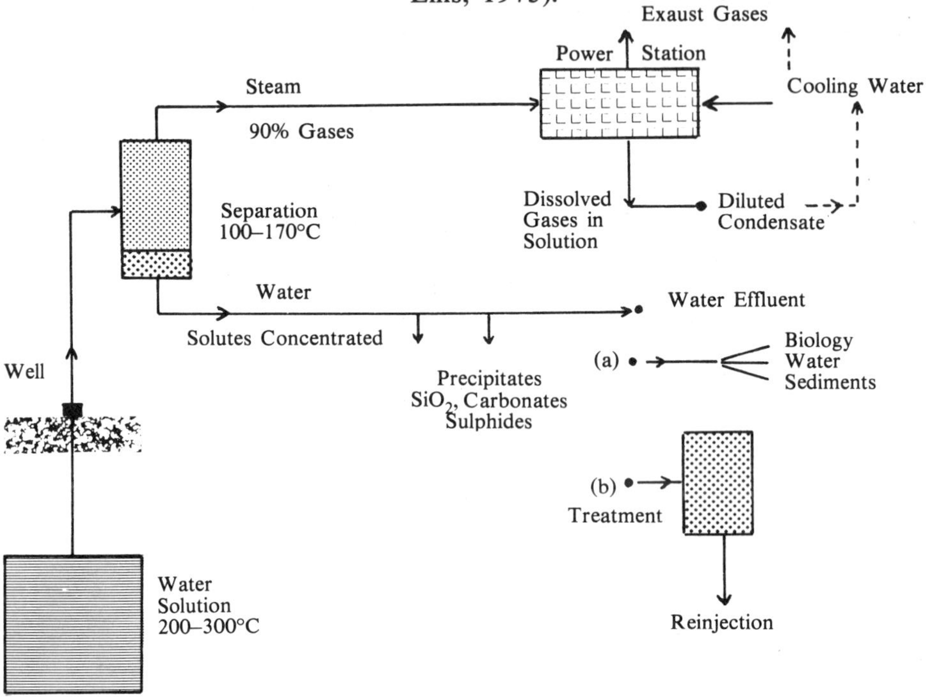

Fig. 2.4 Binary Cycle Process

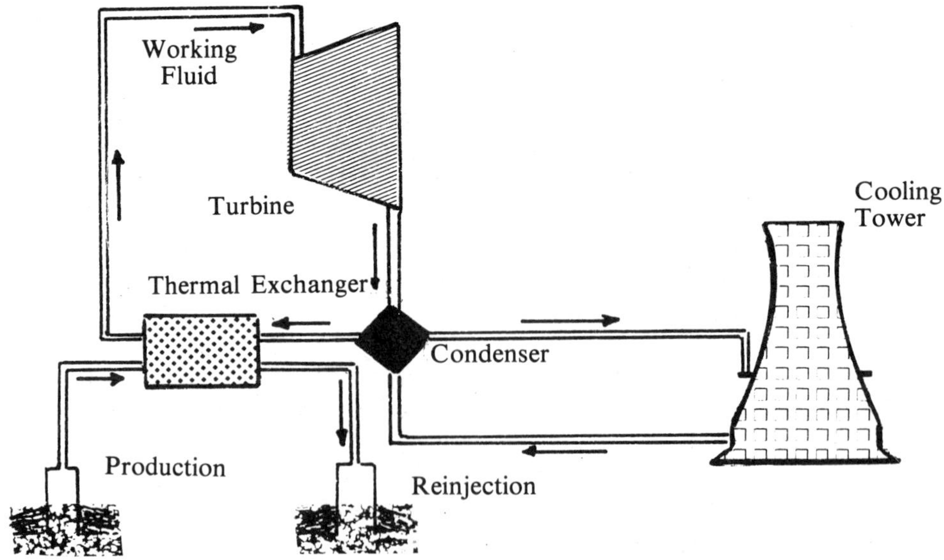

Vapour-dominated systems (also known as dry steam systems) produce superheated steam (about 240°C) at high pressures (up to 35 kg/cm^2) with minor amounts of other gases but little or no water. Since the steam is usually of high quality — that is, it contains few particulates or other substances that must be extracted before use — it can almost be fed directly into a conventional steam turbine to generate electricity. Upon issuing from the turbine, the steam is directed through a condenser where it is condensed; a part of that liquid may be used as a coolant for the condenser and the remainder may be reinjected into the rock formation. Examples of dry steam geothermal fields are Larderello, Italy; The Geysers, California and Matsukawa, Japan.

The second type of geothermal resources (*hot igneous systems*) include both magma and hot impermeable dry rock (the solidified margins around magma and the overlying roof rock). The recovery of geothermal energy directly from magma is not yet feasible. The technology required to utilize hot dry rock is beginning to be developed. Preliminary engineering approaches to tapping the energy potential are focussing on the design of a circulatory fluid flow loop through the rock (Fig. 2.5). First a well would be drilled into the hot formation; then cold water would be injected under high pressure to fracture the formation, and a second well would be drilled to intersect the fractured zone. Finally, surface water would be injected to the first well, passed over the hot dry rock, and withdrawn through the second well in the form of steam or hot water. The heated water generated could then be processed using either the flashed steam or binary cycle process.*

Fig. 2.5 Schematic Diagram of a Hot Dry Rock Heat Recovery Process.

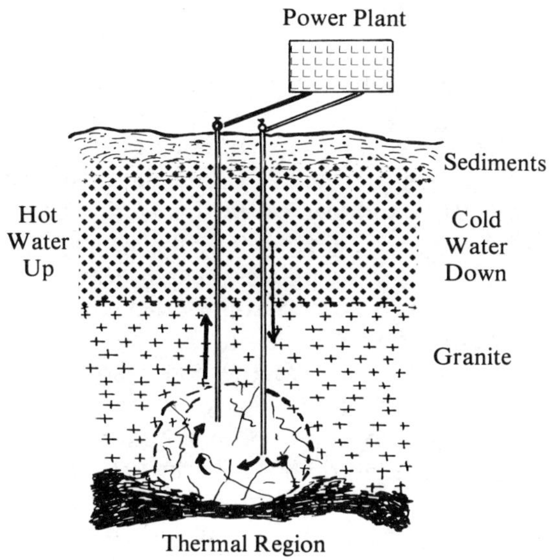

* Advanced testing of this hot, dry rock system is being conducted by the Los Alomos Scientific Laboratory, at their Fenton Hills site, 60 km west of Los Alamos, New Mexico, U.S.A. (Pasqualetti, personal communication).

A third type of geothermal system (Geopressured deposits) arises where a deep sedimentary basin occurs in a zone of high heat flow. In situations such as the Paris Basin or the Hungarian Basin, water may be tapped by wells at temperatures up to about 100°C. A special category of this type occurs in regions where the normal heat flow of the earth is trapped by insulating impermeable clay beds in a rapidly subsiding geosyncline or downward bend of the earth's crust. Geopressured deposits are hotter than normally pressured deposits because upward loss of the included water has been stopped for millions of years. The high temperatures and pressures have resulted in a natural cracking of the petroleum hydrocarbons in the sediment, and in geopressured zones, reservoir fluids commonly contain 1–2 m^3 of natural gas per m^3 of fluid. These dissolved hydrocarbon gases would be a valuable by-product of fluid production. Temperatures of water produced from geopressured geothermal systems would range from 150°C–180°C; well head pressures would range from 280–560 kg/cm^2. Production rates could be several million cubic metres of fluids per day per well and perhaps about 30,000 m^3 of natural gas per day per well (Hickel, 1972). Geopressured reservoirs have been found in many countries while searching for oil and gas; for example in the U.S.A., Mexico, S. America, Far East, Middle East, Africa, Europe and in the U.S.S.R. The exploitation of such reservoirs for geothermal energy has not yet been demonstrated.*

USE OF GEOTHERMAL ENERGY

The utilization of geothermal energy for the production of electricity dates from the early years of the twentieth century. For 50 years the generation of electricity from geothermal energy was confined to Italy (the first experimental electric power generator was operated at Larderello in 1904 and power has been produced there since 1913) and interest in this technology was slow to spread elsewhere. In 1943 the use of geothermal hot water for space heating was pioneered in Iceland. More recently intensive exploration work was undertaken in New Zealand, Japan, the U.S.A. and some other countries, which led to the commissioning of a number of geothermal power stations in these countries. Table 2.1 summarizes the world geothermal installed electrical generating capacity by 1980 and projections to the year 2000. It should be noted that geothermal fields are at various stages of exploration and development in Chile (El Tatio), Nicaragua (Momotombo), Turkey (Kizildere, Afyon), Kenya (Olkaria), Ethiopia, Guadalupe, Western U.S.A., and India. In addition, geothermal exploration is being carried out or considered in more than 50 other countries.

* In the U.S.A., a substantial part of the interest in geopressured reservoirs is in the potential production of natural gas. The heat and mechanical energy of the pressured waters appears secondary (Pasqualetti, personal communication).

Table 2.1 World Geothermal Electrical Generating Capacity (MWe)

Country	Installed capacity	Planned Cumulative (* indicates a minimum figure)			
	1981	1985	1990	1995	2000
Costa Rica	—	—	80	380	380*
Chile	—	—	15	15*	15*
El Salvador	95	150	260	425	535
Ethiopia	—	8	8	8	50
France	—	15	15*	15*	15*
Iceland	32	32	68	68*	68*
Indonesia	0·25	32	92	92*	92*
Italy	440	480	560	620	800
Japan	168	1,000	3,668	3,668*	3,668*
Kenya	—	30	30*	30*	30*
Mexico	150	620	1,000	2,000	4,000
New Zealand	202	202	282	382	382*
Nicaragua	—	—	35	35	100
Philippines	446	558	1,225	1,225*	1,225*
Taiwan	0·4	3	3*	3*	3*
Turkey	0·5	0·5	100	100	150
USA	923	1,674	4,374	4,974	5,824
USSR	5	10	310	310*	310*
Totals	2,472	4,239	8,280*	9,675*	11,714*

Source: United Nations Conference on New and Renewable Sources of Energy, Technical Panel on Geothermal Energy, Final Report Nov. 1980.

The non-electrical applications of geothermal energy are widespread. Early uses included medicinal mineral baths which are well-documented. The use of geothermal energy for space heating has become widespread in many countries; the most significant district heating system is the well-known system of Reykjavik in Iceland. Other countries using geothermal energy for space heating include U.S.A., U.S.S.R., New Zealand, Japan, Hungary, France (Howard, 1975; Banwell, 1975; El-Hinnawi, 1977; Phéline, 1977; Energies Nouvelles, 1978).

Some 25,000 - 30,000 apartments near Paris are heated with geothermal energy and there are plans to extend this system to cover some 500,000 apartments by 1985–1990. Hot geothermal waters are also used extensively in agriculture, especially in greenhouses, in connection with animal husbandry, in aquaculture and in controlled growth of single-cell proteins. Agricultural applications of geothermal energy currently constitute the major part of the non-electrical applications. Over 90 per cent of this capacity is associated with large acreages of greenhouses in the U.S.S.R., where about 25 million m^2 of greenhouses are operated with geothermal waters producing about one million tonne of vegetables per year. There is also a wide range of use of geothermal energy in industry, from the drying of fish, earth and timber, to pulp and paper processing. The two largest present industrial applications are a diatomaceous earth plant in Iceland and a pulp paper and wood processing plant in New Zealand. It should be emphasized that there is a wide range of uses of geothermal fluids for non-electrical purposes, and that several schemes are being developed for combined production power and process heat. Table 2.2 summarizes the total installed geothermal energy for non-electrical use.

Table 2.2 Total Installed Low Temperature (above 15°C) Geothermal Energy, 1980

Country	Installed capacity MW(t)	
Japan	4,475	(81)
Hungary	1,166	(619)
Iceland	1,141	(932)
U.S.S.R.	555	(555)
Italy	~265	(~73)
China	151	(144)
U.S.A.	~115	(111)
France	56	(56)
Czechoslovakia	43	(35)
Romania	36	(36)
Austria	5	(2)
Total	8,008	(2,644)

Figures between brackets indicate uses other than bathing.

Source: United Nations Conference on New and Renewable Sources of Energy, Technical Panel on Geothermal Energy, Final Report, Nov. 1980.

Different estimations have been made for the capacity of geothermal energy most likely to be harnessed by the year 2000. Table 2.3 gives a considered estimate of geothermal capacity that is likely to be harnessed by 1985 and the year 2000; the geological potential of favourable areas is given for comparison.

Table 2.3 Geothermal Energy Scenario until the year 2000

Geothermal energy estimates	1981	1985	2000
I. Geothermal Electricity estimated from collected data [MW(e)]*	2,470	5,000	50,000
II. Geothermal non-electrical energy estimated from collected data [MW(t)]**	2,650	10,000	100,000

* Geothermal Electricity calculated according to geological potential is about 5×10^5 MW(e); see Auer *et al.*, 1978.

** Geothermal non-electrical energy calculated according to geological potential is about 8×10^7 MW(t); see Auer *et al.*, 1978.

Assuming that the total primary energy demand by the year 2000 is 600×10^{18}J (WEC, 1978), geothermal energy that can be harnessed by that time can provide 0·8% of this demand.

ENVIRONMENTAL IMPACTS OF GEOTHERMAL ENERGY

The widespread belief that geothermal resources represent a relatively "clean" non-polluting energy source recently has played an important role in heightening public interest in geothermal development. However, more recent studies (e.g. Bowen, 1973; Axtmann, 1975 a, b; Kruger, 1976; El-Hinnawi, 1977; EPA, 1977; Pasqualetti, 1980) have reviewed the environmental aspects of geothermal energy development and have highlighted a number of potential impacts. Although knowledge of these environmental impacts is still incomplete, geothermal resources do appear to offer several significant environmental advantages over fossil and nuclear energy.

Since geothermal energy must be utilized or converted in the immediate vicinity of the resource, the entire fuel cycle, from resource extraction to transmission, is located at one site. Unlike fossil fuel or nuclear power production, geothermal energy is not a technology that requires a massive infrastructure of facilities and equipment and large amounts of input energy (the fuel cycle for fossil fuel and nuclear power production includes extraction, transportation, fuel processing and waste disposal which are not located at one site but in many cases involve widely separated locations). Furthermore, in some cases, geothermal power does not place excessive demands on water supplies for cooling purposes since the condensed steam could be recycled for that purpose.

However, not all the environmental effects of geothermal energy are positive. There are a number of adverse local effects on land, air and water. These effects are site specific and they vary according to the geochemical properties of the hydrothermal reservoir, the exploitation history of the resource, and according to the carrying capacity of the environment in which geothermal development takes place. Accordingly, generalizations about geothermal emissions are difficult to propound, particularly since the effluent characteristics may be affected strongly by individual plant designs and power cycles (Axtmann, 1975 b). In the present chapter, use is made of the extensive information available for The Geysers, U.S.A. and the Wairakei, New Zealand, geothermal fields as examples of dry steam and hot water dominated fields, respectively. It should be emphasized that the environmental impacts to be encountered in geothermal fields in other sites have to be assessed for each site.

Impacts on Land

The development of geothermal resources has a number of land-use impacts. Prospecting takes place in a wide range of situations, including open farm land, forest, and near closely settled areas, as well as in rough volcanic terrain. The geological, geophysical and geochemical survey work that precedes a geothermal development requires widespread access. Survey lines and vehicle access tracks may have to be opened up in virgin areas. However, the work needs no massive excavations or general clearance. Test drilling in promising areas will require a few access roads for a drilling rig and clearances of about 1000 m^2 for each drilling site.

The intensive development of a geothermal area carries with it the problems of any large civil engineering operation (e.g., road construction, accommodation for workers, and the accumulation of equipment on the site). Patterns of work and ways of life in small communities may be radically changed. Noise from vehicles, diesel-powered drilling rigs, and other mechanical operations can be expected.

As many geothermal areas are associated with soft rocks such as pumices or are in hydrothermally altered rock, development can cause land erosion and slips if care is not exercised. Significant land erosion occurred at Wairakei and at The Geysers (Reed and Campbell, 1975; EPA, 1977), but with experience better techniques have been evolved.

The average overall area involved in a high temperature geothermal development is one of the magnitude of 1 km^2 for 25–50 MWe (Banwell, 1975; Reed and Campbell, 1975). However, the amount of surface land required for the development of a geothermal field differs from one location to another depending mainly on the topography and the well spacing density. From the area presently used at The Geysers (dry steam field), it is estimated that the amount of land actually disturbed per MWe ranges from 0·19 to 0·32 ha (Pasqualetti, 1980).* For a hot water field (Wairakei) the amount of land disturbed is about 1·6 ha/MW (EPA, 1977).

Another important land-use issue is the extent to which such development is compatible with surrounding land uses. Possible adverse effects to adjacent land could result from the changes in the use of the land at the site; e.g. human activity in relation to noise and pollutant emissions. Furthermore, such impacts are likely to be long-term in relation to the life of a geothermal field. Many geothermal developments for electric power production have occurred in passive multipurpose land uses such as wildlife reserves, cattle grazing, and watersheds. At The Geysers, the most visible biological change is the removal of vegetation, and the most serious consequence of such removal is the loss of animal habitat. However, such loss could be minimized if the critical habitat components are preserved (Pasqualetti, 1980). The impacts of geothermal development on land fertility appear to be minimal. During most of the 60 years of field development at Larderello, Italy, for example, the surrounding land has had varied agricultural uses. Today, pipelines traverse vineyards, orchards and farmlands with no known detrimental effect.

* Each 110 MW(e) power plant at The Geysers requires from 14 to 20 wells; each well typically requires from 8–16 ha, but this is bottom-hole spacing. Most wells are now directionally drilled, meaning that a site of less than one hectare can accommodate 5 wells (Pasqualetti, 1980).

Land subsidence resulting from the withdrawal of geothermal fluids from the reservoir is a potential serious impact of geothermal development in some areas. Although land subsidence does not appear to be a problem in the development of dry steam (vapour dominated) geothermal fields, recent studies reported evidence of "significant crustal movement" at The Geysers (Lofgren, 1978). In addition to shifts along the major faults in the region, there have been vertical and horizontal surface deformation over the producing part of the geothermal reservoir system in apparent direct response to fluid withdrawals (see also Pasqualetti, 1980). However, land subsidence in dry fields seems to be very slow and hardly detectable without careful prolonged measurements. In contrast, the problem of land subsidence is more acute for liquid-dominated geothermal fields. In Wairakei, New Zealand, where waste geothermal water is discharged to a river rather than reinjected, total maximum vertical movement has exceeded 3·7 metres since 1956, and has affected an area of 65 km^2, but the effect grades to zero at the perimeter of the affected area (Fig 2.6). Horizontal movement also has been recorded (Axtmann, 1974; Swanberg, 1975; EPA, 1977). Significantly, the area of maximum subsidence may occur outside the production field, which means that subsidence could affect the property of adjacent land-owners more than the immediate development area. In Cerro Prieto, Mexico, a hot water field located near the Imperial Valley in California, subsidence was recorded some 10 km outside the well field even before extensive production began. However, the subsidence is more likely related to regional tectonics and sedimentation than to geothermal fluid withdrawal (Pasqualetti, 1980).

Fig. 2.6 Contours of the Land Subsidence within the Wairakei Field, 1964-1974 (after Stilwell *et al.*, 1975).

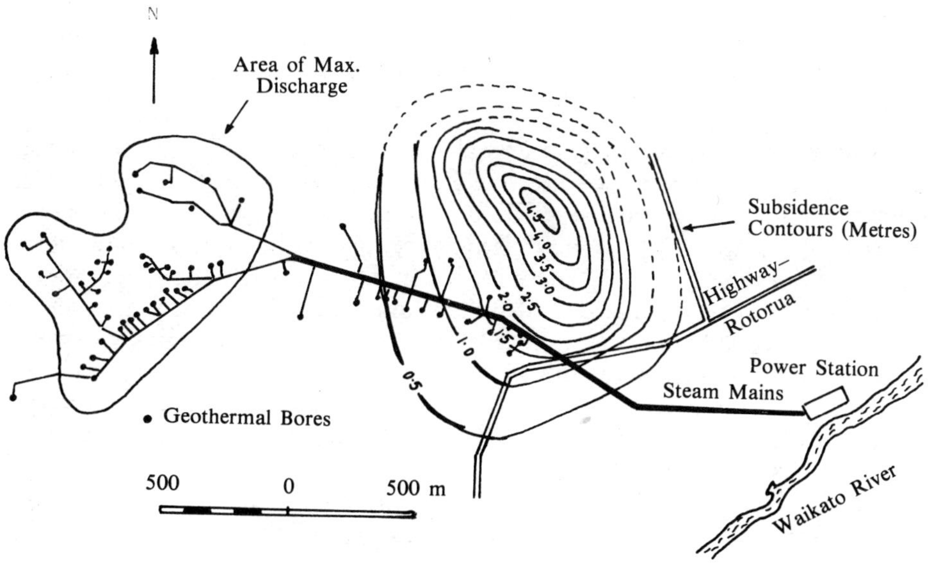

The problem of land subsidence can possibly be alleviated. Reinjection of geothermal fluids to deep wells following power production is highly promising but has some practical limitations. Most geothermal waters contain a large amount of dissolved solids, such as silica or calcium, which may clog reinjection pipes. Such problems could be solved by chemical additives to keep dissolved solids in solution; but their use may create other environmental problems, as well as major costs.

Reinjection may also lower the temperature and hence the energy potential of subsurface geothermal waters (Einarsson *et al.,* 1975). This effect has been observed in Japanese hot water fields. On the other hand, reinjection is potentially an important factor in increasing the productive life of geothermal field (Einarsson *et al.,* 1975), as it is considered a "recycling" of the geothermal waste water through the reservoir. With surface disposal, the residual heat in the water, after separation of the steam, is wasted and its disposal may cause thermal pollution of surface waters. By reinjection this heat energy, which may represent roughly 40 per cent of the heat produced by the well, is also recycled to the reservoir.

The unstable conditions in the earth's crust that create geothermal resources are conditions that also produce seismic activity. In fact, geothermal resources currently being developed are located in seismically active zones and micro-earthquakes (less than 4 on the Richter scale) have been observed near many major geothermal areas around the world. It has been mentioned that reinjection of fluids into tectonically active areas induces seismic activity (Swanberg, 1975). Although this has been the case with deep well disposal of other liquid wastes, reinjection experiments in geothermal fields in Italy have produced no seismic or micro-seismic effects (El-Hinnawi, 1977). Monitoring at The Geysers, showed no evidence to indicate that geothermal development has increased the seismicity of the area (EPA, 1977).

Noise

The development of a geothermal field creates considerable noise, particularly at the drilling and well-testing stages. Drilling operations, with noise from diesel engines and other heavy equipment, can create localized noise levels of 80–90 dBA for 24 hours a day. However, this problem is not unique to geothermal development and is encountered in other construction operations. The noise from the first discharging of wells is intense and can create annoyance at distances of several kilometres. Suter (1978) surveyed the effects of geothermal development on wildlife behaviour, while Andersen (1975) noted the effect of noise on farm animals. Unsilenced geothermal wells may produce noise levels of up to 120 decibels in their near vicinity, and to prevent damage to hearing, workers must wear ear protectors.

Once cleared from drilling mud and rock cuttings, wells can be discharged through efficient silencers that reduce the noise level to below 100 dBA near the well, and at a distance, the sound is reduced to a low rumbling which as noted by Armstead (1975) has a soporific rather than a distressing effect. In addition to lowering the overall noise, the silencers remove the high frequencies that are typical of a geothermal well blasting steam and water into the atmosphere at a supersonic speed. By reducing the time of first blowing of wells to an absolute minimum and with the early installation of efficient silencers, geothermal noise can be reduced and is not likely to be a major environmental problem.

In rare situations, ground vibration may be produced within about a hundred metre radius from a well that has got out of control during drilling or through a break in the casing. As pointed out by Jhaveri (1975) continuous low-frequency vibration can cause irritability and lowered work performance. "Rogue" wells are a temporary phenomenon and are either quenched by a secondary drilling operation or may with time tend to quench themselves by rock slumping. The local ground vibration created by a "rogue" well at Wairakei became a popular tourist curiosity while it persisted for a few years.

Airborne Effluents

The major sources of air pollutants emitted during geothermal power production are direct releases of geothermal steam during all stages of development and releases of non-condensable gases during plant operation. The pollutants likely to result from geothermal power development are primarily determined by the chemical composition of the deep geothermal fluid at a site. Both the total quantity of gases in the fluid and the relative concentration of their constituents depend on the geochemistry of the underground reservoir.

The compositions of geothermal steam in various areas are summarized in Table 2.4. In steam-dominated fields (The Geysers and Larderello), the discharged steam composition corresponds to that at depth. However, in high temperature water-dominated fields the proportion of gas in the steam depends on the extent to which steam has flashed from the original high-temperature water. The gases (except ammonia) are predominantly concentrated in the steam phase and the gas/steam ratio decreases with increasing steam proportion in the discharge. The gas contents vary widely, and occasionally beyond the range in Table 2.4. For example, at Ngawha, New Zealand and at Monte Amiata, Italy, wells discharged steam containing up to 50 per cent of gas (mainly CO_2), but fields were not major producers (Ellis and Mahon, 1977). In this high gas situation the H_2S/CO_2 ratio is usually low. Carbon dioxide is usually the predominant gas, but the proportions of other gases vary widely. In extreme cases in active volcanic areas, e.g., Namafjall, Iceland, and Tatun, Northern Taiwan, gases may contain 10–30 % hydrogen sulphide. There are other areas with extremely low H_2S concentrations in steam, as at Salton Sea, California, and Ahuachapan, El Salvador. In the fields so far developed, in both water- and steam-dominated areas, changes with time in the gas content of the steam flow tend to be negative, so that the initial assessment gives the most pessimistic picture of the pollution potential (Ellis and Mahon, 1977).

Table 2.4 Concentration of Gases in Geothermal Steam (mg/kg)

Area	Separation Pressure (bars. abs.)	CO_2	H_2S	NH_3	CH_4	H_2	N_2	H_3BO_3	Ref.
Larderello (average)	·	47,500	540	150	215	50	240	220	(1)
The Geysers (average)	·	3,260	222	194	194	56	52	91	(2)
Cerro Prieto (average)	10·5	14,100	1,500	110	·	·	·	<0·1	(3)
Wairakei (average)	10·9	2,300	72	4·8	2·5	0·5	17	<0·1	(4)
Broadlands Well 8	5·8	30,000	350	34	420	3	230	<0·1	(4)
Otake Well 7	3·0	2,400	16	← Remainder 67 →	← →	← →	← →	← →	(5)
Hveragerdi Well 4	1	1,210	36	·	1·6	2·5	·	·	(6)
Salton Sea (average)	1	3,000	2·6	35	10	·	·	·	(7)

References: (1) ENEL (1970); (2) Reed and Campbell (1975); (3) Mercado (1976 A); (4) Ellis (1978); (5) Noguchi (1966); (6) Arnorsson (1974a); (7) Gudiksen, *et al.*, (1978).

Table 2.5 gives the amounts of the main airborne effluents for some geothermal plants. A 100 MWe plant in a steam-dominated area requires about 10^7 tonnes of geothermal steam per year; in water-dominated fields this steam would be accompanied by between 2 and 10 times the amount of hot water.

Table 2.5 Main airborne effluents [tonne per 100 MWe(y)]

Location	H_2S	NH_3	CO_2
The Geysers	2,000	1,700	30,000
Larderello	5,000	1,300	400,000
Wairakei	300	50	8,000
Broadlands	6,000	500	400,000
Cerro Prieto	4,000	800	150,000

After Ellis (1978).

Hydrogen sulphide is the main airborne effluent of real concern at present. Standards for air quality vary, but in populated areas the limits for H_2S in air are often close to the limit of detection for the human nose; by volume 0·002 ppm (Axtmann, 1975a). Although H_2S is a poisonous gas at high concentrations, the main problem in geothermal development is simply its objectionable smell. However, traces in the air may also create problems with electrical equipment, cause general corrosion of metals, and blacken lead-based paints. Sub-ppm levels do not appear to present a health problem, and communities have lived for generations with local air containing 0·01–0·1 ppm (e.g., Rotorua, New Zealand).

As H_2S is a heavy gas it tends to concentrate around geothermal plant in enclosed areas, in well-head pits and in underground cable-ways. H_2S becomes a health hazard at concentrations of over 10 ppm. At Cerro Prieto there was trouble with hydrogen sulphide wafting to the ground near the plant from the 40m gas stacks, and fans and ducting were installed to remove the gas from the vicinity (Mercado, 1975).

Hydrogen sulphide passing through a power station with direct condensers distributes itself between the gas ejectors, the cooling tower plume and the reject condensate in a manner which varies with the design and operation of the plant. In the early Geysers power stations for example, about equal quantities of H_2S were emitted through each route (Bowen, 1973). The atmospheric disposal of waste gases was considered satisfactory in the early days of the geothermal development at The Geysers, but when H_2S emissions rose from 14 kg/hour in 1960 to 750 kg/hour in 1975, widespread complaints were received from nearby Lake County residents (Leibowitz, 1977). In the Wairakei station about 30% of the inflow H_2S exits from a 30 m gas exhaust chimney and the dispersal of the gas leads to concentrations of up to 1 ppm H_2S in the air within a 0·5 km radius. There has been little comment from the nearest permanent residents (0·8 km). The remaining H_2S enters the river water/condensate, and together with the reject hot water causes the odour of H_2S to persist in the Waikato river for a distance downstream.

Various processes have been proposed for limiting H_2S emission from geothermal stations, and several have been tried in the The Geysers plants (Laszlo, 1976). These include several modifications of the Claus reaction (Allen and McCluer, 1975; Axtmann, 1975b). The Jefferson Lake modification appears to be the most likely of this type, and Velker and Axtmann (1978) consider that recovery of sulphur may give break-even economics in the U.S.A. where plants emit more than about 20 tonnes per day of H_2S. The "iron catalyst" modification is in operation on existing Geysers units with efficiencies of about 40–70%. It involves the gases pumped from direct condensers being burnt to form SO_2, and the scrubbing of the gases by cooling tower water to which an iron catalyst has been added. It is not altogether satisfactory because of corrosion and sludge formation in the plant. At The Geysers the Stretford process is now favoured, and it is being installed in the newest unit. Shell and tube stainless steel condensers provide a highly concentrated gas steam to the process which involves washing the gases with a solution of sodium carbonate, sodium ammonium polyvanadate and anthroquinone disulphonic acid. The hydrogen sulphide becomes oxidized to sulphur by the vanadate and the reduced vanadium is restored to vanadate by air. High-quality sulphur is recovered. The economics of operation are not yet established, but Laszlo (1976) considered that the process may raise the cost of the power plant by 45%. The level of current control technology suggests that hydrogen sulphide abatement levels of over 90% are possible.

Recent work by Thompson and Kats (1978) showed that constant H_2S concentrations of 0·3 ppm or over affected the foliage and growth of California buckeye, lettuce, alfalfa, sugar beet, grapes, ponderosa pine, and Douglas fir.

Rapid growing crops such as grapes, alfalfa and lettuce were more affected than the slower-growing buckeye and pine. Douglas fir was more sensitive than ponderosa pine. Low concentrations of H_2S at 0·1 and 0·03 ppm had little effect, and actually stimulated the growth of the field crops. These effects would probably be the maximum to occur, as field concentrations would be highly variable according to wind conditions. A practical example is provided by the Rotorua, New Zealand, public gardens where a wide range of healthy trees and flowers has existed for over 100 years within a few hundred metres of thermal activity causing local H_2S concentrations of about 0·1 ppm. Hydrogen sulphide oxidises slowly to sulphur dioxide over a period of about a day. By this time the sulphur gases would be well dispersed into the atmosphere. Radiata pine shows good tolerance to foliar concentrations of sulphate at many hundreds of ppm, but the long-term effects of consistent SO_2 air concentrations at the 0·01 ppm and of the possibility of formation of acid rains near geothermal plants need further investigation.

Other gaseous effluents include carbon dioxide and ammonia. Carbon dioxide mainly enters the exhaust gas stream and does not constitute an environmental problem near the geothermal plant, although it adds to the global atmospheric CO_2 budget (Reed and Campbell, 1975). However, at Wairakei, where river water is the condenser water supply, 30–35% of the carbon dioxide passes into the reject condensate and then to the Waikato river where it may contribute to increased weed growth. Ammonia emitted in the exhaust gas stream is rapidly diffused by atmospheric processes and the values are rapidly lowered to acceptable values. However, if ammonia reacts with other chemicals (such as hydrogen sulphide to form ammonium sulphate), harmful environmental impacts on human health and certain plant and animal species may result (EPA, 1977).

Radon-222 is found in trace amounts in the non-condensable gas portion of geothermal steam. While the values vary widely, both between and within fields, they are in the range 3–200 nCi/kg of steam (see Table 2.6). The levels of radon measured in the ambient air outside the geothermal plant at The Geysers are in the range of 0·03–0·15 pCi/l* (Reed and Campbell, 1975, Stoker and Kruger, 1975), well below the average for air over continental land areas which is about 0·3 pCi/l (Stoker and Kruger, 1975).**

Several geothermal areas are situated in districts with known mercury mineralization (e.g., The Geysers, California and Ngawha, New Zealand) and higher, than normal, regional concentrations of mercury occur in the air (Siegel and Siegel, 1975). In geothermal systems, mercury concentrates with great preference into the steam and gas phases rather than into the hot water. The mercury is present mainly as elemental mercury in spite of the presence of sulphur (Robertson *et al.*, 1978). Table 2.7 gives the concentrations of mercury in geothermal steam of various areas. In a power station most of the mercury

* pCi = pico Curie = 10^{-12} Ci.

** More recently, Anspangh (1978) indicated that the maximum rate of release of radon from the entire field is 1·43 Ci/day and that its effect is not discernible because of the naturally occurring levels of Rn-222 and its daughters in the environment.

Table 2.6 Concentration of Rn–222 in Geothermal Steam

Location	nCi/kg of steam*	Ci/100 MW(e)y**
The Geysers	20	180 (1)
Larderello	20–200	180–1,800 (2)
Wairakei	10–100 (variable)	~900 (3)
Broadlands	1–5	9–45 (3)

Source: Reed and Campbell, 1975, Stoker and Kruger, 1975; D'Amore (1975); Dall'Aglio (1976); Ellis (1978).

*n = nano, i.e. 10^{-9}.

**Assuming that 10^7 tonnes of steam are required to produce 100 MW(e)y and that 90% of the radon enters the exhaust vapour.

Table 2.7 Concentration of Mercury in Geothermal Steam

Location	*μg/kg in steam*	*kg Hg/100 MW(e)y*
The Geysers	5	45 (1)
Matsukawa	50–60	450–540 (2)
Wairakei	1–4	9–36 (2)
Broadlands	4–8	36–72 (2)
Salton Sea	10–20	95–180 (3)
Cerro Prieto	7–9	63–81 (3)
Namafjall	· 0·2	1·8 (4)
Otake	0·3–2	2·7–18 (2)

References: Reed and Campbell (1975); Ellis (1978); Robertson *et al.*, (1978); Olafsson (1977).

from well discharges passes through the turbines into the condensers. Between 10 and 50% of the mercury is dissolved in the condensate; the remainder passes into the exhaust gas. At Cerro Prieto, for example, 6·8 g/hour is discharged into the air of a total 7·4 g/hour throughout. With good dispersal into the atmosphere, e.g., by injecting the waste gases into the cooling tower plume, it can be calculated that mercury concentrations in the immediate vicinity of the station would be about 0·01–0·1 $\mu g/m^3$ (Ellis, 1976). These levels are unlikely to present a health hazard. (Although Schroeder (1971) considered that prolonged exposure to atmospheric levels of mercury in excess of 0·1 $\mu g/m^3$ may be harmful, the adopted maximum level for 8-hour industrial exposure is 50 $\mu g/m^3$ (Sax, 1975)).

Liquid Effluents

In steam-dominated systems (e.g. at The Geysers), the steam after passing through the turbines is generally condensed by contact with cooling water. Cooling water consists normally of steam condensate which has passed through an open cooling tower cycle. Since the amounts of steam condensate are much higher than the needs for cooling, a substantial fraction has to be disposed of as liquid effluent. A 100 MWe power station at The Geysers produces about 3800 m³/day of waste water, the composition of which is given in Table 2.8.

Table 2.8 Composition of Condensate return from a 100 MWe Geothermal Plant (The Geysers)

Composition	*Tonne/y*
Nitrate	0·15
Magnesium	1·5
Sulphate	197·0
Chloride	5·3
Silica	5·5
Calcium	8·0
Sulphur (free)	13·3
Boron	25·9
Ammonia	223·5
Organics	311·9
Alkalinity (HCO_3)	648·0

After EPA (1977).

In liquid-dominated systems the mixed steam/water discharge from a well passes at a controlled pressure into cyclone separators. Gases ex-solved from the original deep high-temperature water concentrate mainly in the steam phase (CO_2 + H_2S + minor gases), while the water becomes slightly concentrated through steam loss, and rises in pH through the loss of acidic gases. The separated water phase is discharged through silencers (open towers acting as atmospheric pressure separators), emerging as an effluent at about 100° (current practice at Wairakei). Alternatively, the separated water, still under pressure, may be disposed of into reinjection wells (as practiced at Ahuachapan and several Japanese stations) or under the surface of a disposal pond (as at Cerro Prieto). The water mass flows are considerable (approximately twice to 10 times the steam flow). The overall efficiency in converting the thermal energy of well discharges to electricity is about 7–10%, lower than for the steam-dominated areas (about 15%). This neglects the possible secondary utilization of separated hot water for heating purposes, or the return to the aquifer of the heat contained in the reinjected water.

A hot water geothermal electric station therefore may have four major effluent streams: hot saline water, surplus condensate retaining some dissolved gases, a waste gas flow, and impure vapour rising from cooling towers. The chemical constituents present in the deep water are distributed finally between each flow in a manner related to their relative solubilities and fractionation kinetics.

The quality of the geothermal water — its physical and chemical characteristics — varies widely. While some geothermal hot waters contain relatively few pollutants, most contain a relatively large amount of dissolved solids (and heavy metals) because the high temperatures of the brines increase the dissolution rate of solids and heavy metals in the rock. Table 2.9 gives the analyses of waters in some high-temperature areas associated with power development. The most common type of geothermal water is a sodium, potassium chloride solution containing appreciable concentrations of silica, boron, sulphur, ammonia, fluoride, and various trace metals* (see also Table 2.10). Although most geothermal waters have common features, particular areas may be associated with high concentrations of specific elements. As examples, the El Tatio field in Chile has waters containing up to 45 mg/kg arsenic and Ngawha, New Zealand contains up to 1000 mg/kg boron. It should also be noted that medium- and low-temperature geothermal waters used for non-electrical purposes vary widely in composition depending upon the geological setting of the reservoir and its geochemistry (see Table 2.11). In this case, the possible pollution effects of each "waste" water need to be assessed individually.

Table 2.12 gives the outflow of liquid effluents from some hot-water geothermal power stations.

Based on the variability in the amount and type of dissolved solids in geothermal fluids, a number of different methods for disposing of wastewater have been tested and used. These include direct release to surface water bodies, evaporation, surface spreading to shallow aquifers, desalination with subsequent water re-use, and re-injection to the production reservoir. A common problem is that of dissolved silica. Silica in the waste water commonly precipitates as amorphous silica in the waste water channels, in which case the channels must be cleaned at regular intervals (Fig. 2.7). The selection of a disposal method depends on local hydrologic conditions, water requirements, the quality of waste water and environmental regulations.

For most power units in water-dominated geothermal fields, direct discharge of waste water is unlikely to be satisfactory. For example, there is concern in New Zealand about the long-term effects on the Waikato River from the waste waters discharged from the Wairakei plants. Coulter (1978) noted that nitrogen is the factor limiting phytoplankton growth in the Waikato River in summer, and that further input from the Broadlands station would be undesirable. Weissberg and Zobel (1973) showed that mercury concentrations varied up to about 1 mg/kg in bottom sediments of the Waikato River following the entry of effluent waters from several geothermal areas. Fish in parts of the river had average mercury concentrations ranging up to about 0·8 mg/kg, i.e. above the accepted 0·5 mg/kg level for human consumption. Arsenic discharge from the wells in the reduced trivalent form is rapidly oxidized to the pentavalent form when the waste waters are diluted by river. Although some arsenic is taken up by weed growth in waterways (Reay, 1972), its behaviour downstream is predominantly that of a soluble element. Further studies on the transport mechanism, fate and effects of

* It should be noted that detailed studies of the possible patterns of migration of trace elements during the exploitation of geothermal fields should be undertaken to determine the path of such elements and their possible environmental impacts (Brondi et al., 1980).

Table 2.9. Compositions of some high-temperature geothermal waters (concentrations in ppm in waters collected at atmospheric pressure)

Source	Aprox Temp. (°C)	pH	Li	Na	K	Cs	Mg	Ca	F	Cl	Br	I	SO_4	HCO_3	B	SiO_2	H_2S	NH_3	P	Ref.
Well 24, Wairakei, N.Z.	250	8·3	13·2	1,250	210	2·5	0·04	12	8·4	2,210	5·5	0·3	28	24	29	670	1	0·2	0·01	(1)
Well 8, Broadlands, N.Z.	260	7·9	11	975	232	1·6	0·1	3	8·4	1,858	6·3	0·6	3·5	157	53	796	2	2·3	—	(1)
Well 9, Otake, Japan	215	8·15	5·2	936	131	0·8	0·19	12	4·7	1,474	3·4	0·26	136	47	20	310	—	0·06	0·3	(2)
Well 15A, Mexicali, Mexico	250	—	15	6,000	1,125	—	—	321	2·2	11,500	—	—	0	43	9	678	—	22	—	(3)
Well 7, El Tatio, Chile	255	7·3	48	5,000	840	18	0·1	203	2·5	9,100	—	—	29	44	210	810	—	3·1	—	(1)
Well 1A, Kizildere, Turkey	200	9·0	4·5	1,280	135	0·3	0·1	3·0	24	117	1·2	—	770	1,350	26	263	—	2·5	—	(1)
Well 1, Ngawha, N.Z.	230	7·4	12	950	77	0·4	—	28	0·85	1,625	—	—	17	86	1,200	460	—	46	—	(1)
Well 8, Reykajanes, Iceland	290	7·1	6·6	12,730	1,990	—	9·8	2,249	0·14	25,054	87	0·27	2·4	—	11	943	—	2·0	—	(4)
Well 205, Matsao, Taiwan	245	2·4	26	5,490	900	9·6	131	1,470	7	13,400	—	—	350	—	106	639	—	36	—	(1)
Well IID–1, Salton Sea, Ca.*	340	4·7	215	50,400	17,500	14	54	28,000	15	155,000	120	18	5	7,100†	390	400	16	386	—	(5)

References: (1) Ellis (1978); (2) Koga (1970); (3) Mercado (1976 B); (4) Olafsson and Riley (1978); (5) Muffler and White (1969).

* Concentrations and pH in deep aquifer.

† Co_2 concentrations.

Table 2.10. Trace elements (μg/kg) concentrations in geothermal well waters

Element	Fe	Mn	Cu	Pb	Zn	Cd	Tl	As	Sb
Wairakei, N.Z.	12	0·7	2	4	2	0·05	—	4,700	100
Broadlands, N.Z.	360	13	1	4	1	0·02	7	3,300	200
Mexicali, Mexico	200	640	5	5	6	—	—	2,000	400
Matsao, Taiwan*	148,000	28,000	35	500	8,800	—	—	—	—
Salton Sea, California	$2{\cdot}3 \times 10^6$	$1{\cdot}4 \times 10^6$	8,000	1×10^5	5×10^5	2,000	1,500	12,000	400
Reykjanes, Iceland	·	2,600	14	·	92	1·1	·	146	0·6

References: (1) Ellis (1978); (2) Mercado (1967); (3) White (1968); (4) Olafsson and Riley (1978).

* A very low pH water (pH 2·9).

Table 2.11. Hot waters representative of types used for community or agricultural heating (Concentrations in ppm)

Source	Temp. (°C)	pH	Li	Na	K	Ca	Mg	Fe	F	Cl	Br	I	So_4	HCO_3	B	SiO_2	H_2S	NH_3	P	NO_3	Ref.
Rotorua. Well 231	110	8·9	1·8	345	17	1·5	0·5	—	6·9	343	1·0	0·5	22	194	5·1	274	37	0·2	—	—	(1)
Boise. Idaho	75	—	—	75	1·3	2	—	—	24	9	—	—	23	141	—	78	—	—	0·01	0·08	(2)
Well 4. Szentes. Hungary	85	—	—	378	18	7·8	—	1·4	2·8	13	—	0·03	13	1,049	0·3	55	—	6·2	—	—	(3)
Spring. Carlsbad Czechoslovakia	73	7·65	3·3	1.718	104	102	47	0·1	2·4	617	1·4	—	1.662	2.100	1·0	68	—	0·1	—	—	(4)
Laugarás. Iceland	98	8·8	—	79	2·1	4·1	0·01	—	2·1	56	—	—	62	18	0·32	127	0·7	—	—	—	(5)
Well 1. Tigre Lagoon. Louisana (3930 m)	—	5·4	—	43.000	290	1.660	161	10	0·95	67.100	65	25	200	645	60	63	1	90	—	—	(6)
Well. Kings Country. California	100	5·7	15	13.600	404	12.200	275	0·1	1·6	44.000	238	56	16	80	50	63	0·6	127	0·1	0	(7)

References: (1) Glover (1967); (2) Kunze, *et al.*, (1975); (3) Boldizar (1975); (4) Ovchinnikov (1955); (5) Arnorsson (1974b); (6) Hankins *et al.*, (1978); (7) White (1965).

Table 2.12. Outflow of liquid effluents per 100 MW(e)y from three hotwater geothermal sites (tonne/y)

Constituent	Cerro Prieto	Wairakei	Broadlands
Li	320	350	300
Na	150,000	35,000	30,000
K	40,000	5,500	5,000
Ca	10,000	500	80
F	40	200	200
Cl	300,000	60,000	50,000
Br	500	150	150
SO_4	170	700	200
NH_4	800	50	500
B	400	750	1,100
SiO_2	40,000	20,000	20,000
As	30	100	100
Hg	—	0·025	0·035

Fig. 2.7 A view of the Main Wairakei Water Disposal Drain Showing Heavy Silica Encrustations. The Silica in the foreground has been removed mechanically, revealing the thick deposits of Silica on the drain lower margins.

chemical elements in waste waters directly discharged to surface waters in other geothermal fields are needed.

Ponding of water effluents to evaporate and to allow the possible recovery of dissolved salts, presents interesting possibilities in dry climates. Ponding also lowers the possibility of thermal pollution of receiving waters. At Cerro Prieto, an 8 km^2 pond allows the concentration of the 2000 tonne/hour water flow by a factor of 2–3, with some precipitation of silica and heavy metals, before the effluent enters the Rio Hardy (Mercado, 1977). The recovery of salts is also being examined. Fernelius (1975) described the operation of a desalination plant with geothermal brines at East Mesa, California. In arid areas where irrigation and drinking water is valuable, an integrated desalination, a salt recovery/water supply industry may be usefully associated with geothermal power production. Such a scheme was proposed for the El Tatio, Chile, geothermal area.

Moderate temperature waters (up to 130°C) used for various domestic, commercial and agricultural heating purposes do not present unique problems and should be considered as any other waste water whose environmental significance is related to its salinity and chemical compositions. However, large volumes of water are involved in heating schemes, e.g. a 1 MWt heating system based on an 85°C water source requires a water flow of about 18 tonnes/hour. As an example, in Iceland where a community of 110,000 about Reykjavik receives heat from wells tapping a dilute water at about 100°, there have been no disposal problems from 9,000 tonnes/hour of water at 40°C being discharged to waste. The waters are little different in composition from the local surface waters (Arnorsson, 1975). In arid areas, dilute thermal waters may be valuable in supplementing local water supplies. Hydrogen sulphide is often at a low concentration level and can be removed by oxidation.

Reinjection is the favoured means of waste water disposal for most geothermal developments. However, it cannot be described yet as well-established technology, except for the simple case of steam condensate reinjection where relatively small volumes of water are involved, and the solutions are dilute. In an area where water supplies are drawn from underground aquifers it is important to have a knowledge of the local hydrology and to carefully monitor any effects arising from the injected geothermal waters. In high-temperature water areas it is likely that emergency discharge ponds would need to be provided in case reinjection wells ceased to operate temporarily. For example, at Broadlands, New Zealand, a 40 hectare pond 5 m deep is planned to store 30 days discharge from the proposed station (NZED, 1977). Reinjection has been practiced satisfactorily to date at Ahuachapan, The Geysers, Larderello-Travale, but well scaling and thermal degradation problems have arisen at Otake and Hatchobaru, Japan.

Thermal Discharges

Geothermal power plants have a low thermal efficiency; for dry steam operated plant the efficiency is about 15% whereas for liquid-dominated geothermal field the efficiency is about 8–10%. In other words about 85 to 92% of the total heat energy contained in the geothermal fluid is emitted as waste heat. The considerable outflow of water vapour from the silencers, condensate trap valves, hot-water drains, and from the cooling towers of a power station, may in some climates create a tendency for local fog formation and icing of roads in winter. An average 110 MWe Geysers station cooling tower emits 658 tonnes of water vapour per hour (Reed and Campbell, 1975). The problem is greater for stations with mechanical draught cooling towers than for natural draught cooling towers which have a greater plume rise. The 900 MWe installation at The Geysers could increase the moisture content in the air of the local basin by more than 50% (Swanberg, 1975).

Water-dominated fields have an appreciable potential for thermal pollution. In future developments this need not be the case since reinjection is becoming common, and the currently-operating stations mark only stages in the development of a new technology. For example, the low efficiency of the Wairakei station is partly due to an early concept of combining power generation and heavy water production. Axtmann (1975a) reviewed the extreme case of the Wairakei station where once-through river water cooling is used and separated hot water also enters the Waikato River. Of the 1720 MW of energy piped to the surface, some 850 MW finally enter the river through condensate (580 MW) and waste hot water (270 MW). At the average river flow this raises the river water temperature by about 1·5°C after mixing, but at very low flows temperature rises of several degrees can occur and there is evidence of the fish life being affected (Coulter, 1978). In few situations will river water cooling be permissible, as temperature rises of a few degrees are disastrous for many kinds of fish. For example, trout usually cannot survive above about 23·5°C and the optimum conditions are 13–16°C. Their eggs do not hatch above 14·4°C (McKee and Wolf, 1971). In hot water fields other than Wairakei, stations use a cooling tower/condenser cycle. Combined with waste hot water reinjection (or ponding, as practiced at Cerro Prieto) hazards to local waterways from thermal pollution can be minimized. Far better would be the utilization of separated hot water (115–160°C) for community or industrial heating, before reinjection.

References

Allen, G. W. and McCluer, H. K. (1975): Abatement of hydrogen sulphide emissions from The Geysers geothermal power plant. Proc. 2nd U.N. Symp. Develop. Use Geothermal Resources, San Francisco, 2, 1313–1321.

Andersen, S. O. (1975): Environmental impacts of geothermal resource development on commercial agriculture. Proc. 2nd U.N.Symp. Develop. Use Geothermal Resources, San Francisco, 2, 1317–1321.

Anspaugh, L. R. (1978): Final Report on the Investigation of the Impact of the Release of Rn-222, its Daughters, and Precursors at The Geysers Geothermal Field and Surrounding Area. Lawrence Livernmore Lab., Environmental Science Division.

Armstead, H. C. H. (1975): Environmental factors and waste disposal. Proc. 2nd U.N.Symp. Develop. Use Geothermal Resources, San Francisco, 1, lxxxvii.

Arnorsson, S. (1974a): The ulitity of waters from high-temperature areas for space heating. Geothermics 3, 127–141.

Arnorsson, S. (1974b): Composition of thermal fluids in Iceland. In "Geodynamics of Iceland and the North Atlantic Area", D. Reidel Publ. Co., Dordrecht, Holland.

Arnorsson, S. (1975): Geothermal energy in Iceland: utilization and environmental problems. Naturopa, No. 23, 23–26.

Auer, P. L. *et al.* (1978): Unconventional Energy Resources. In World Energy Resources 1985–2020. World Energy Conference. IPC Science Tech. Press.

Axtmann, R. C. (1974): An environmental study of the Wairakei power plant. N.Z. Dept. Sci. and Ind. Res. Phys. Eng.Lab. Rept. No. 445.

Axtmann, R. C. (1975a): Environmental impact of a geothermal power plant. Science 187, 795–803.

Axtmann, R. C. (1975b): Emission control of gas effluents from geothermal power plants. Environ. Letters 8, 135–146.

Banwell, C. J. (1975): Geothermal energy and its uses: Technical, economic, environmental and legal aspects. Proc. 2nd U.N. Symp. Develop. Use Geothermal Resources, San Francisco, 3, 2257–2267.

Boldizar, T. (1975): Research and development of geothermal energy production in Hungary. Geothermics 4, 44–56.

Bowen, R. G. (1973): Environmental impact of geothermal development. In "Geothermal Energy", eds. P. Kruger and C. Otte, Stanford University Press.

Brondi, M. *et al.* (1980): Stable and Radioactive Trace Elements in Hydrothermal Fluids from the Campi Flegrei Area, Naples, Italy. Proc. 3rd Inter. Symposium on Wate-Rock Interaction, Edmonton, Canada, p. 192.

Coulter, G. W. (1978): A preliminary appreciation of effects on aquatic environments of geothermal power development in New Zealand. Geothermal Res. Council Trans. 2, 119–120.

Dall'Aglio, M. (1976): l'impatto ambientale dell'energia geotermica con particolare riferimento alla radioattivita. Simpasio Intern. Sobre Energia Geotermica en America Latina, 761.

D'Amore, F. (1975): Radon-222 survey in Larderello geothermal field. Italy. Geothermics 4, 96–108.

Einarsson, S. *et al.* (1975): Disposal of Geothermal Water by Reinjection. Proc. 2nd U.N. Symp. Develop. Use Geothermal Resources, San Francisco. 2, 1349.

El-Hinnawi, E. E. (1977): Energy Environment and Development 10th World Energy Conference. Istanbul 1977 paper 4. 8–1.

Ellis, A. J. (1975): Some geochemical problems in the utilization of geothermal waters. Proc. Grenoble Symp., Int. Assn. Hydrol. Sci., Publ. 119, p. 100–109.

Ellis, A. J. (1976): Geothermal fluid chemistry and human health. Simp. Int. Sobre Energia Geotermica en America Latina, Guatemala City.

Ellis, A. J. and Mahon, W. A. J. (1977): "Chemistry and Geothermal Systems", Academic Press. New York: 392 pp.

Ellis, A. J. (1978): Environmental Impact of Geothermal Development. Report prepared for UNEP. October 1978.

ENEL (1970): Larderello and Monte Amiata: Electric Power by Endogenous Steam. Ente Nazionale per l'energia Elettrica, Rome.

Energies Nouvelles (1978): La Géothermie en France. Collection Energies Nouvelles; B.R.G.M. Orléans, Février.

EPA (1977): Western Energy Resources and The Environment: Geothermal Energy. EPA-600/9-77-010 Washington, D.C.

Fernelius, W. A. (1975): Production of salt water by desalting geothermal brines. Proc. 2nd U.N. Symp. Develop. Use Geothermal Resources 3, 2201–2208.

Glover, R. B. (1967): The chemistry of thermal waters at Rotorua. N.Z.J. Sci. 10, 60–96.

Gudiksen, P. H. *et al.* (1978): Air quality assessment studies of geothermal development in the Imperial Valley. Geothermal Res. Council Trans. 2, 235–236.

Hankins, B. E. *et al.* (1978): Chemical analysis of water from the world's first geopressurized geothermal well. Geothermal Res. Council Trans. 2, 253–256.

Hickel, W. J. (1972): Geothermal Energy. Univ. of Alaska.

Howard, J. H. (1975): Principal conclusions of CCMS non-electrical applications project. Proc. 2nd U.N. Symp. Develop. Use Geothermal Resources, San Francisco, 3, 2127–2141.

Jhaveri, A. G. (1975): Environmental noise and vibration control at geothermal sites. Proc. 2nd U.N. Symp. Develop. Use Geothermal Resources, San Francisco, 2, 1375–1378.

Koga, A. (1970): Geochemistry of the waters discharged from drillholes in the Otake and Hatchobaru areas. Geothermics Sp. Issue, 2,2, 1422–1425.

Kruger, P. (1976): Geothermal Energy. In Annual Review of Energy Vol. 1 p. 159. Annual Reviews Inc. Palo Alto, California.

Kunze, J. F. *et al.* (1975): Non-electric utilization projects, Boise, Idaho. Proc. 2nd U.N.Symp. Develop. Use Geothermal Energy, San Francisco 3, 2141–2145.

Laszlo, J. (1976): Application of the stretford Process for H_2S abatement of The Geysers geothermal power plant. 1976 Intersociety Energy Corrosion Conf. (A.I.Ch.E., HT and EC Div.), 7 pp.

Leibowitz, L. P. (1977): Projections of future H_2S emissions and geothermal power generation: The Geysers region, California. Geothermal Res. Council Trans. 1, 183–185.

Lofgren, B. E. (1978): Monitoring crustal deformation in The Geysers — clear Lake geothermal area, California. U.S. Geological Survey. Open-File Re. 78–597, Sacramento.

McKee, J. E. and Wolf, H. W. (1971): "Water Quality Criteria". California State Water Resources Control Board, Publ. 3A.

Mercado, S. (1967): Geoquimica Hidrotermal en Cerro Prieto, B.C. Mexico. Comision Federal de Electricidad, Mexicali.

Mercado, S. (1975): Cerro Prieto geothermoelectric project: pollution and basic protection. Proc. 2nd U.N. Symp. Develop. Use Geothermal Resources, San Francisco 2, 1394–1398.

Mercado, S. (1976a): Cerro Prieto geothermal field, Mexico: wells and plant operation. Proc. Int. Congress on Thermal Waters, Geothermal Energy and Volcanism of Mediterranean Area. 1, 394–408, National Technical University, Athens.

Mercado, S. (1976b): Disposicion de desechos geotermicos. Simp. Int. Sobre Energia Geotermica en America Latina, Guatemala City.

Mercado, S. (1977): Salmuera de desecho de la planta geotermica de Cerro Prieto. Problematica y posibles beneficios.

Muffler, L. J. P. and White, D. E. (1969): Active metamorphism of Upper Cenozoic sediments in the Salton Sea geothermal field, Geol. Soc. Am. Bull. 80, 157–182.

Noguchi, T. (1966): Exploration of Otake steam field. Bull. Volcanol, 29, 529–544.

NZED (1977): "Broadlands Geothermal Power Development Environmental Impact Report". New Zealand Electricity Dept. Wellington.

Olafsson, J. (1977): Mercury and arsenic in Namafjall and Krafla drillholes. Jokull.

Olafsson, J. and Riley, J. P. (1978): Geochemical studies on the thermal brine from Reykjanes, Iceland. Chemical Geology 21, 219–237.

Ovchinnikiv, A. M. (1955): Mineral waters of the health resort Karlovy Vary in Czechoslovakia. Vopr. Kurortol. Fiz. Lecheb. 3, 66. (Chem. Abstr. 52, 2311).

Pasqualetti, M. J. (1980): Geothermal Energy and The Environment: the Global Experience. Energy, 5, 111.

Peters, W. *et al.* (1978): An appraisal of World Coal Resources. In World Energy Resources 1985–2020. World Energy Conference. IPC Sci. Tech. Press.

Phéline, J. (1977): Développement et Utilisation des Sources Alternatives d'Energie. Revue Générale Thermique, No. 185.

Phéline, J. (1979): Energie Solaire et production décentralisée d'electricité. Annales Des Mines, April 1979.

Reed, M. J. and Campbell, G. E. (1975): Environmental impact of development in The Geysers geothermal field, U.S.A. Proc, 2nd U.N. Symp. Develop. Use Geothermal Resources, San Francisco, 2, 1399–1410.

Reay, P. F. (1972): The accumulation of arsenic from arsenic-rich natural waters by aquatic plants. J. Appl Ecology 9, 557–565.

Robertson, D. E. *et al.* (1978): Chemical characterisation of gases and volatile heavy metals in geothermal effluents. Geothermal Res. Council Trans. 2, 579–582.

Sax, N. I. (1975): "Dangerous Properties of Industrial Materials" (4th Ed.), Van Nostrand Reinhold Co., New York.

Schroeder, H. (1971): "Air Quality Monograph". Amer. Petroleum Institute, Washington.

Siegel, S. M. and Siegel, B. Z. (1975): Geothermal hazards: mercury emission. Environ. Sci. Technol. 9, 473–474.

Stilwell, W. B. *et al.* (1975): Proc. Second U.N. Symposium Development and Use of Geothermal Energy, San Francisco, 2, p. 1428.

Stoker, A. K. and Kruger, P. (1975): Radon in geothermal reservoirs. Proc. 2nd U.N. Symp. Develop. Use Geothermal Resources, San Francisco 3, 1797–1803.

Suter, G. W. (1978): Geothermal development and wildlife behaviour. Geothermal Res. Council Trans. 2, 627–629.

Swanberg, C. A. (1975): Physical aspects of pollution related to geothermal energy development. Proc. 2nd U.N. Symp. Develop. Use Geothermal Resources, San Francisco 2, 1435–1443.

Thompson, C. R. and Kats, G. (1978): Effects of continuous H_2S fumigation on crop and forest plants. Environ. Sci. and Technol. 12, 550–553.

Velker, J. A. and Axtmann, R. C. (1978): Sulfur emission control for geothermal power plants. M/S in preparation. Princeton University, Chemical Engineering Dept., Princeton, New Jersey.

WEC (1978): World Energy Resources 1985–2020. World Energy Conference, IPC Sci. Tech. Press.

Weissberg, B. G., and Zobel, M.G.R. (1973): Geothermal mercury pollution in New Zealand. Bull. Environ. Contamination and Toxicol. 9, 148–155.

White, D. E. (1965): Geothermal Energy. US Geol. Survey Circular 519.

White, D. E. (1968): Environments of generation of some base-metal ore deposits. Economic Geology 63, 301–355.

CHAPTER 3

Solar Energy

JEAN PHÉLINE
Adviser on alternative energy sources
Commissariat A l'Energie Atomique, Paris, France

Fig. 3 Example of collective application of solar water heating, at Cagnes sur Mer, South of France (Doc. COMES)

INTRODUCTION

Despite the explosive growth in worldwide energy demand, full awareness of the limits of non-renewable energy sources and their environmental impacts was brought into focus only in the 1970s following the 1973 oil crisis. National and international actions have begun to harness renewable sources of energy, particularly solar energy; interest in which had previously been confined to a few far-seeing pioneers.

The relationship between solar energy and the environment is at the very foundation of all biological life on our planet. Of all sources of energy, it is the energy transported to the earth by solar radiation that supports life through photosynthesis and respiration, which are central processes in the metabolism of plants and animals.

This chapter describes the current state of the art in solar technology and the prospects for tapping the vast resources of solar energy, then discusses the different environmental impacts of solar energy exploitation. Only direct utilization of solar energy is considered. Indirect uses, for example, wind power, ocean thermal energy conversion (OTEC), hydropower, wave and tidal energy, and biomass energy are discussed in other chapters.

SOLAR RADIATION

The energy radiated from the sun has its origin in the thermonuclear reactions within the star. The quantities involved are stupendous: every second, the sun's mass diminishes by about 5 million tonnes, all converted to radiation. Only a small proportion (about one ten-billionth) of this energy reaches the earth. This minute fraction is, however, equivalent to more than 150,000 million megawatts, which is over 10,000 times the world's total installed generating capacity. It has been estimated that the solar energy received each year on earth (around 10^{18} kWh) exceeds 20,000 times the world's total annual energy consumption (UNEP, 1979).

However, all the solar energy received on the earth's surface cannot be recovered. A large part is used up by natural processes such as ocean water evaporation, global water circulation (clouds, rainfall, streams and rivers, etc.), winds, and growth of plants and animals. In addition, it is the sun's energy that maintains temperatures suitable for human and other biological life.

The solar energy received at ground level varies between 0 and 1,100 watts per square metre of horizontal surface, according to latitude and season. This variation is due to many phenomena, some regular and predictable (alternation of day and night, seasonal variations, etc.), and others essentially random (atmospheric impurities, cloud cover, and so on).

The solar radiation reaching the earth's surface consists of about 3% ultraviolet, 42% visible and 55% infrared, and is of three types:

- *direct radiation*, received directly from the sun, when the sky is clear. Before entering the atmosphere, its energy varies between 1,350 and 1,450 W/m², depending on the distance of the earth's surface from the sun;
- *diffuse radiation*, resulting from reflections in the atmosphere and clouds. This radiation reaches the ground from all directions (as opposed to direct radiation), and its energy is mainly a function of cloud cover;
- *reflected radiation*, consisting of radiation reflected from the earth's surface (such as soil, vegetation, snow and water, and buildings). Its energy depends on the surface's albedo (i.e., the fraction of light reflected by the surface).

Meteorologists regularly use sunshine recorders to measure the time (in hours and tenths of hours) during which the solar disc is clearly visible at a given location, based on a brightness threshold corresponding to an insolation of about 100 W/m² (Perrin de Brichambault and Lambolley, 1975). Engineers and designers performing heat balance calculations use published formulas and charts to convert these measured sunshine values into figures for total radiation and energy received on surfaces with selected orientations and tilts. The Global Meteorological Satellite System (GMSS) and the European Heliosat programme use meteorological satellites to make pictures of the earth, with sufficiently close areal coverage and at short time intervals, to estimate global irradiation levels. However, the calculation of radiation balances of the earth and its atmosphere is highly complex, due to the variability of the main influencing factors. For most practical applications, use is, therefore, made of the daily, monthly, seasonal or regional averages of measurements at ground-level meteorological stations of the total solar radiation on horizontal and vertical surfaces (Fig. 3.1). More detailed information for particular sites is obtained with portable data acquisition and recording systems (Colomes and Roland, 1977).

Fig. 3.1 Daily Global Irradiation Levels in kWhk/m² (July average).

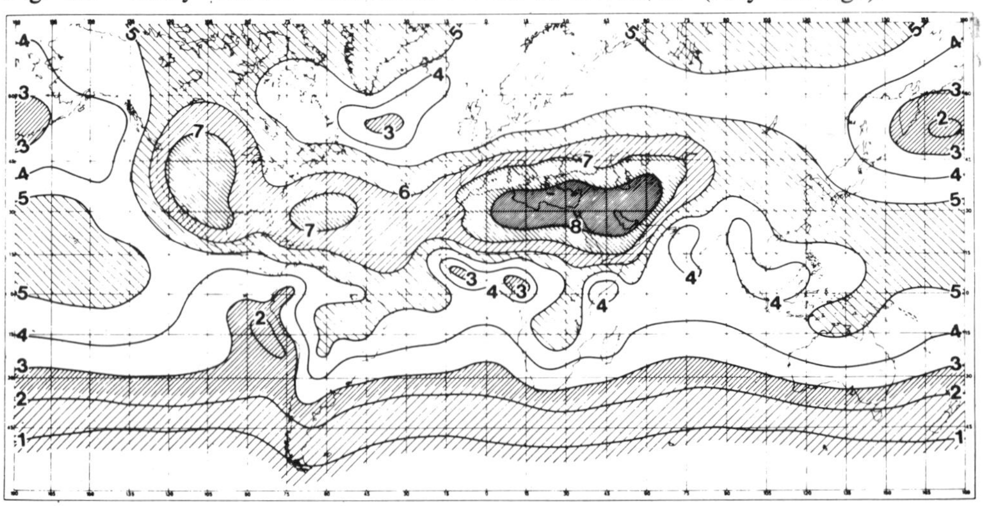

After: French National Meteorological Institute.

The extraction of all the solar energy reaching the earth's surface is nearly impossible because of the diffuse nature of the energy over an extremely large area (the entire surface of the planet), and the periodic and random aspects of its reception at any one place. Harnessing large quantities of solar energy on earth requires either relatively few but very large receiving units, such as fields of concentrating or non-concentrating collectors, or numerous small individual receivers, such as solar houses, heaters, etc. However, these restrictions do not apply if solar satellite systems are introduced. By the turn of this century solar power satellites employing comparatively small collection surfaces to beam enormous quantities of energy to earth may be put into orbit. The only currently feasible energy-collection technology for such satellites is the photovoltaic cell, whereas many widely differing processes (including photovoltaics) are used in earth-based solar energy systems.

DIRECT USES OF SOLAR ENERGY

Flat-Plate Collectors and Water Heating

Fig. 3.2 illustrates a typical flat-plate solar collector. This type of collector is used for low-temperature heating applications, particularly for water heating and space heating. Many widely differing collector designs are available on the market, ranging from very sophisticated selective surface types, which can raise the temperature of the heat-transfer fluid (water, air, etc.) from 40–60 °C to 80–120 °C, to much simpler models with lower performance. All collectors are, however, based on the same physical principles: solar radiation capture by an absorber heat-exchanger and the greenhouse effect (Peri, 1977).

The solar water heater is the best-known and commonest application of the flat-plate collector (solar space heating is discussed in the following section). It consists of one or several collectors, a means for storing the heated water, and control systems. There are two main classes of solar water heater systems:

(*a*) systems with hot water storage in the collectors themselves. This is the simplest method, and is very common in some countries (e.g. in Japan). In general, the hot water is used at the end of the day, to avoid losing the stored heat by convection and radiation at night.

(*b*) systems with separate hot water storage: – thermosyphon systems, in which the heat-transfer liquid circulation is exclusively by natural convection under the influence of the sun, and no mechanical energy is required.

– circulating pump systems which have the highest efficiency, and are most common for large installations. However, mechanical energy is necessary (for pumping), and maintenance is more frequent than with other systems.

Fig. 3.2 Flat-plate Collector.

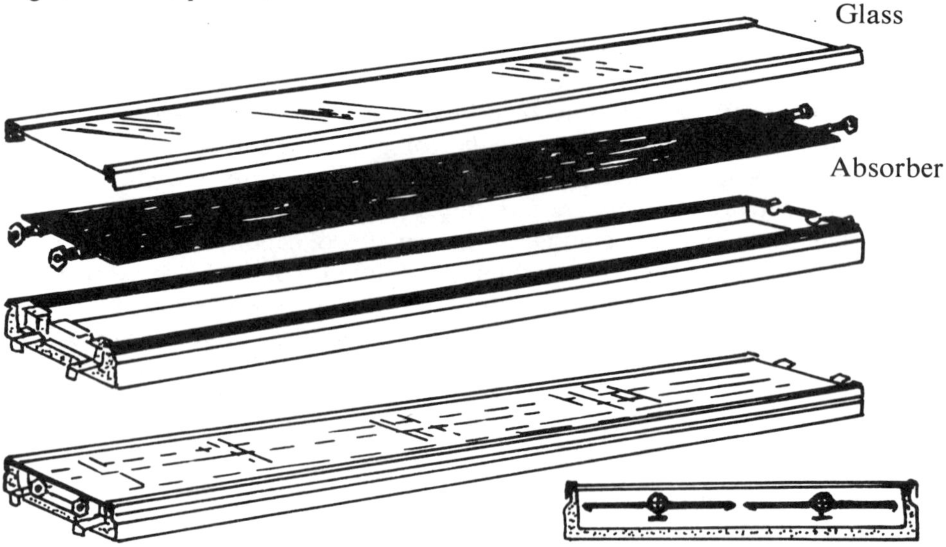

In addition, very simple models of solar water heaters made of plastic are used for heating swimming pools.

A correctly installed solar water heating system has a capacity of about 300 to 500 kWh per square metre of collector surface per year, depending on the geographical location. Auxiliary heating by oil, gas or electricity is often provided for periods of insufficient sunshine and exceptionally cold weather.

Several countries have been using solar water heaters for many years. Around 2 million units are in service in Japan; they are mandatory by law in the Northern Territory of Australia, and around 30,000 units exist in this country (Morse, 1977); approximately 20% of homes in Israel have solar water heaters. Serious production has started in the United States, the United Kingdom, the Federal Republic of Germany, France, Niger, Mali, Senegal and several other countries. Many governments are actively promoting and funding demonstration systems for collective buildings like schools, hospitals and office blocks. Regulations have been issued to make sure that design and installation of solar water heating systems satisfy health and safety requirements (e.g., protection against contamination of tap water by toxic products used as anti-freeze additives, etc.).

The cost of solar water heaters varies greatly, according to design, materials used, and quality. A premium metal-glass collector system with insulated hot water tank usually involves a collector price of about $100 per square metre in the United States. The total installed cost of a typical family-size installation having 3 m^2 of collector surface and a 200 litre storage tank is then roughly $ 600–800. The simpler type of system, in widespread use in Japan, with the hot water stored in the collectors themselves and available only at the end of the day, costs around $ 100–150.

The simplest solar water heaters have no moving parts at all, and all types are essentially trouble-free plumbing systems with a normal service life of 10 years or more. Maintenance is minimal, depending largely on the rate of corrosion, fouling and formation of encrustations due to impurities in the water.

Solar water heaters not only satisfy a fraction of the total energy needs in both developed and developing countries, but they also have beneficial environmental effects, especially in rural areas of developing countries. They are certainly the most promising immediate application of solar energy, leading to significant savings in fossil fuels, and in many cases they have a positive impact on local employment, labour skills and rural development in general.

Space Heating

Application of solar energy to space heating is increasing quite rapidly, mainly in countries with temperate climates, where residential energy consumption constitutes a large proportion of total low-temperature energy use.

Two different systems of space heating are employed:

- *active systems*, in which flat-plate or other types of collectors replace conventional heat sources (furnaces, electric heaters, etc.),
- *passive systems,* whereby homes and other buildings are designed and oriented in an optimal manner with respect to the sun and local weather conditions. The term "bioclimatic architecture" has been used to describe this approach.

In both cases, buildings are carefully insulated ("weatherized") to reduce heat losses.

Active solar space heating

In active systems, collectors (usually flat-plate) capture solar energy when available and transfer it to a heat distribution or storage facility. An auxiliary heating system of the conventional type (electric, oil, etc.) is generally provided as backup for times such as exceptionally cold spells when the solar system cannot supply part or all of the necessary heat (see Fig. 3.3).

Unfortunately, the heat storage capacity of active solar systems is presently limited to just a few hours or days of sunshine, and this is a serious limitation to their widespread use. To overcome this difficulty, interseasonal storage methods are being developed, by which solar heat can be stored for 3 to 6 months, and then recovered when needed (e.g., in winter). Candidate methods include:

- hot water storage in large burried tanks (Torrenti *et al.,* 1976),
- heat storage in shallow groundwater, and
- a range of processes making use of physical and chemical phenomena: heat storage in high heat-capacity materials such as crushed stone and gravel; storage using the latent heat of phase change or reaction in

hydrated salts (at low temperature), in mineral products like soda and potash, in molten salts (nitrates, etc.), in molten metals, and so on (Chaulliaguet, 1979).

Fig. 3.3 Solar Heating System with Oil or Gas Heating Backup system.

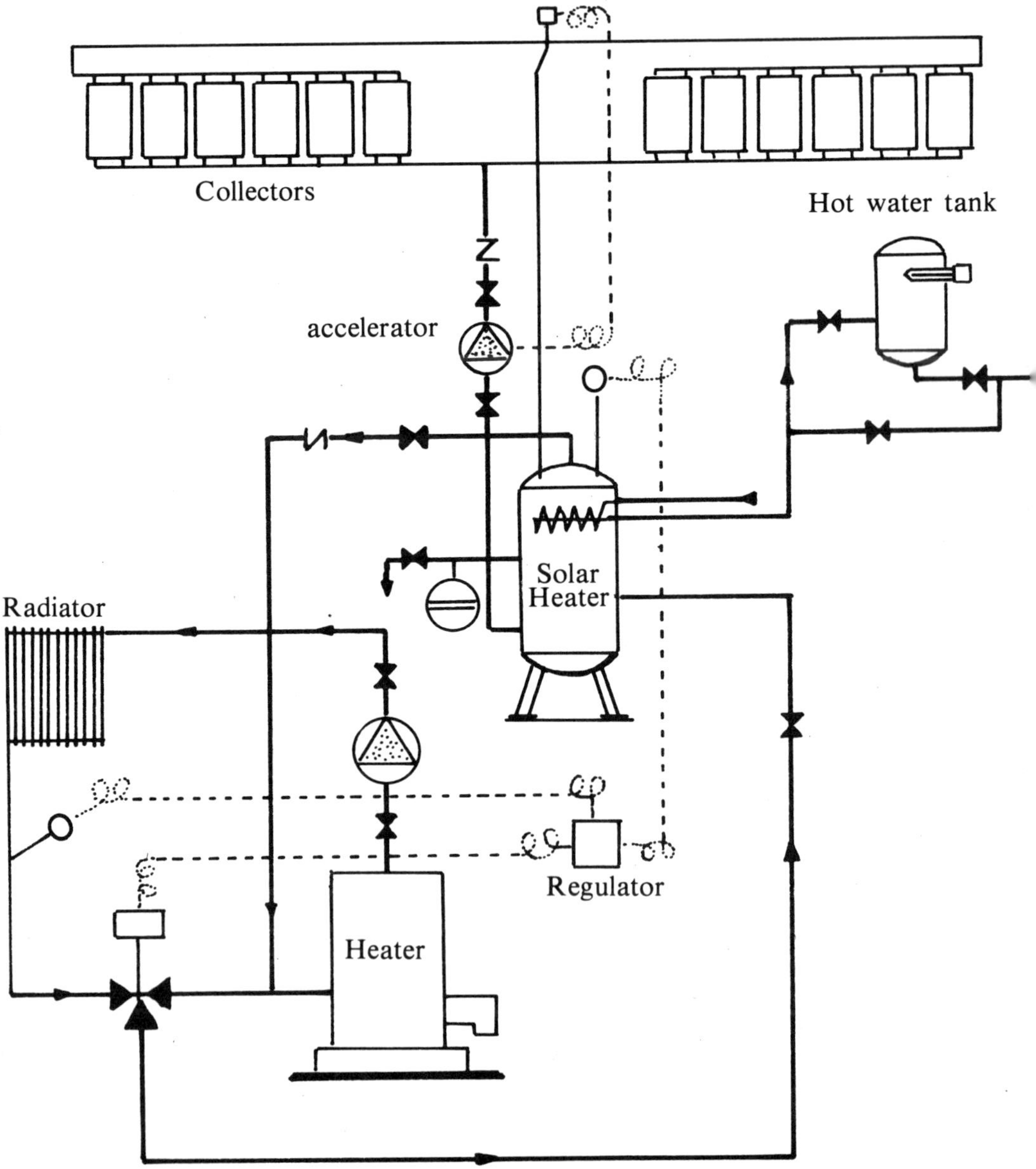

A further difficulty is the fact that the solar collector efficiency decreases with the increase in heat-transfer fluid temperature. The system operating temperature must be as low as possible in comparison to the high temperatures for heat distribution (Gicquel, 1978). One solution being tried out in several countries is to use heat pumps in association with the solar collectors, but it is too early to judge the economic feasibility of such an approach.

Another variant is the trickle-down collector, in which water runs down by gravity between two plates, and is then returned to the top of the collector by an electric pump with low energy consumption. Flow through the collector is in bursts several minutes long; the collector is normally empty of water outside periods of solar energy capture (e.g., at night). The fact that special precautions need not to be taken against freezing in winter avoids the addition of anti-freeze products to the water (usually toxic compounds). This will markedly eliminate possible health hazards at the thermal storage level (Liebard, 1980).

Passive solar space heating

The passive solar space heating approach is being vigorously pursued in many countries, in parallel with the rapid development of active techniques. In this approach, buildings are designed and oriented in an optimal manner with respect to the sun's position and the local climate, to limit or even obviate the need for artificial heating. In many cases, traditional local building techniques are used, as well as more recent modern ideas based on the properties of construction material and understanding of heat exchange phenomena.

One of the commonest new passive techniques is the Trombe wall, named after its French inventor (Trombe and Michel, 1977). This is basically a quite thick, black-surfaced, south-facing wall, usually of concrete, with glazing just in front of it. The sun's radiation penetrating the glazing is trapped behind it, and heats the wall. This stored heat is distributed throughout the house/building by radiation and natural circulation of air between the wall and glazing. Shutters or insulated curtains are sometimes used to reduce heat losses at night.

Heating requirements are also reduced by increasing the amount of insulation. The combination of very high levels of insulation and passive solar techniques has led to the so-called "zero energy" house. Without going so far, more conventionally designed homes can be kept cooler in summer and heating bills can be reduced in winter by simple architectural additions like roof overhangs, conservatories and verandas, etc.

A great deal of research is underway in many countries to develop construction materials and methods for better self-regulation of the temperature in passive solar homes. For example, temperature control can be improved by using materials with high latent heats of fusion-solidification as large-capacity isothermal heat stores (60 to 80 kWh/m^3). Another promising line of investigation is on special membranes with absorption, transmission and reflection properties which vary according to their solar exposure: these could, for example, allow passage of solar radiation when heating is needed, but block it at other times.

Prospects

Although, at first sight, there are two competing solar space heating methods (active and passive), things are not quite so simple in practice. Active solar homes are also designed to make maximum use of the "free" energy of the sun, and purely passive systems usually have imperfect temperature control. In a great many cases, the better approach is therefore a hybrid one drawing on the maximum benefits from passive concepts, but also employing some active components (solar water heaters, motors, controls, etc.). The solar houses of the future will probably be of this type (Chouleur, 1975).

Design and calculation methods have been widely published for application by house builders and architects in optimizing solar heating systems as a function of local climatic conditions (Alexandrof and Liebard, 1979; Presente, 1977). This should help overcome the present hesitation of owners, builders and planners who remain attached to "traditional" building methods.

Several countries are consenting major efforts towards the promotion of solar heating installations as part of their national programmes to reduce dependence on conventional sources of energy (usually imported). A good example is the U.S.A.'s target of 2.5 million solar heating systems by 1985. These efforts are generating a wealth of new designs and modified original architectures; and hundreds of experimental and demonstration solar houses have already been built throughout the world.

In the past, one of the main limitations to widespread use of solar heating was its high initial cost compared to the lifetime savings in fuel. Solar energy could not compete in the industrialized world with conventional heating techniques based on fuels such as oil, natural gas and liquified petroleum gas (LPG), although electric heating was in some cases as expensive as solar heating. However, this situation has changed radically during the last few years. The prices of of oil and gas have climbed so steeply that solar energy for heating homes and collective buildings is already competitive with these fuels, or will become so in the very near future.

Space Cooling, Air Conditioning and Refrigeration

In hot, sunny regions of the earth, the prime requirements is often for space cooling and air conditioning, rather than heating, in which case solar energy can be used as the heat source for the cooling system. Several techniques are available which include:

(*a*) *absorption and adsorption systems*, which are the most common, consisting of two main types:
   - systems using an ordinary liquid absorption cycle, with water-ammonia or water-lithium bromide usually chosen as the absorber-refrigerant mixture. The solar source temperature must be high enough to obtain a good coefficient of performance (see Fig. 3.4);

- systems using a solid absorption or adsorption medium. In the latter case, the adsorber (e.g., zeolite) has a very tight lattice for good adherence of the molecules of the refrigerant fluid (typically alcohol), which are later desorbed under the influence of solar heat;

(*b*) *conventional compression systems*, in which the compressor is driven by a solar thermal or photovoltaic motor;

(*c*) *evaporative cooling systems*, where air is circulated over a wet medium, giving up heat to evaporate the water. Such systems have been experimented in the United States and Australia;

(*d*) *cooling based on nocturnal radiation to the sky*, where a heat-transfer fluid is circulated through a heat exchanger such as a rooftop basin of water which is uncovered at night and covered during the day.

Fig. 3.4 Space Heating, Space Cooling and Water Heating system of The Solar House Designed by the Instituto Tecnológico y de Estudios Superiores de Monterrey (ITESM) in Mexico.

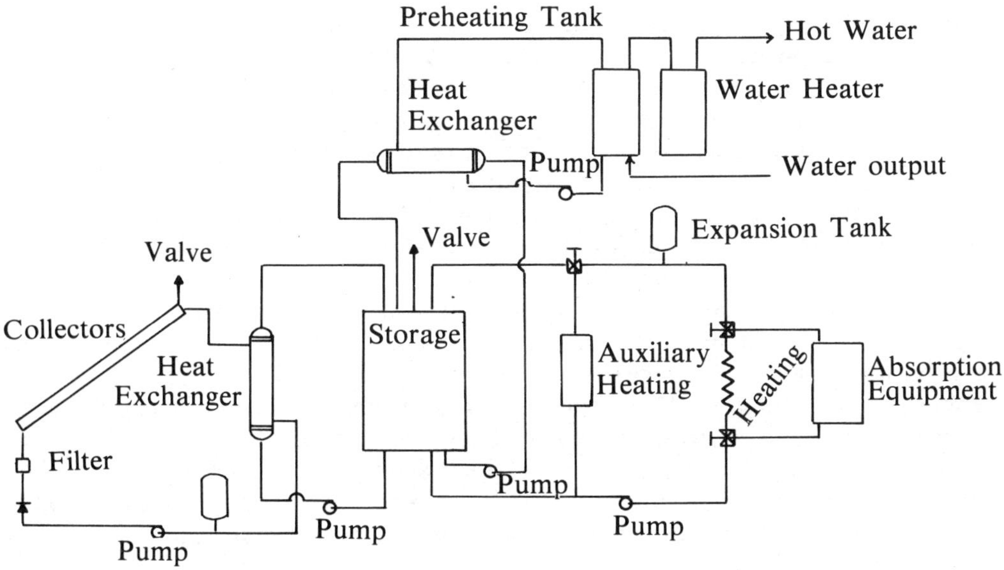

Relatively few experimental solar space cooling systems have been built to date. Although generally successful, they indicate rather high costs compared to standard compression cooling (NAS, 1976). Most have involved supplying a solar-heated absorber-refrigerant mixture. Required operating temperatures for standard absorption systems are obtainable with the more efficient flat-plate collectors and with evacuated tubular collectors; therefore an ordinary commercial air conditioner can easily be used in a solar cooling system, with the solar-heated fluid substituting for the normal heat source (Walon, 1979).

Solar refrigeration is closely related to solar space cooling and air conditioning. Its main potential applications are food preservation and storage of biological and medical products. Experiments are being conducted in several countries, including the United States, Sri Lanka, France and the U.S.S.R., based on absorption cooling cycles and conventional compression cooling cycles using solar-powered motors. A good example is a 25 m^3 Aswan cold store for fish (storage capacity 4 tonnes) built in Egypt with French assistance. However, most development work aims at household refrigerators/freezers and small-scale production of ice.

Decentralized Electric Power Production

Solar energy being available everywhere on earth, even though intermittent, variable and unevenly distributed, is a promising natural energy source for decentralized production of electricity for domestic and collective uses, particularly in small communities (Palz, 1977; Durand, 1979; Bloss *et al.,* 1979). The annual sunshine varies in Europe between 1,450 and 2,800 hours, with the highest values close to the Mediterranean; it exceeds 3,000 hours in many developing countries, attaining 3,600 hours on the high plateaux of Bolivia, in Chili, Iran and Sudan. The corresponding solar energy, expressed as the mean total daily radiation on a horizontal surface at ground level, ranges from 2 to 3 kWh/m^2 in Europe and 5 to 6 kWh/m^2 in most developing countries to over 7 kWh/m^2 in Saudi Arabia, Australia and Peru.

Whereas industrialized countries have dense power transmission and distribution networks reaching out to even the smallest villages, this is not the case in most developing countries. Moreover, transportation networks in the latter countries are often inadequate for the transport of fossil fuels necessary for power production at sites far from urban centres. Considering the current status of solar technology and the present high costs of solar equipment, the first major applications of solar electricity will, therefore, probably be in arid regions and remote communities with good sunshine conditions. With some possible exceptions (Australia, parts of the United States), the introduction of solar electricity in industrialized countries is subject to techno-economic competitiveness with more conventional methods of power production.

SOLAR ELECTRICITY TECHNOLOGIES

At present, there are three main methods of converting solar radiation into electricity (Bloss *et al.,* 1979):

1. *photovoltaic conversion*, in which solar energy is converted directly to direct current (DC) electricity by solar cells. These are diode-like devices

made of semiconducting material. Photons in the light received by each cell release their energy to produce excess charge carriers (electrons and holes) in the semi-conductor material. These carriers cross the diode potential barrier, and thus set up an electric current that can be extracted from the cell. The efficiency obtainable in practice with the usual single-crystal silicon cells is about 10%. Conversion efficiencies of 15 to 18% are possible with different materials and more advanced techniques,

2. *low-temperature thermal conversion*, making use of both direct and diffuse solar radiation; and
3. *high-temperature thermal conversion*, making use of direct solar radiation only.

Both thermal methods convert solar radiation into electricity in two stages: solar radiation is first converted to thermal energy (heat) by a solar collector system; this heat is then converted to electricity with a heat engine. The low-temperature and high-temperature methods differ essentially by the use of solar radiation concentration in the latter systems, by the heat-transfer medium employed and above all its temperature, and by the thermodynamic cycle employed. Overall conversion efficiency ranges from just a few percent with some low-temperature flat-plate collector systems to almost 20% at the high operating temperatures that can be obtained with concentrating collectors in high-temperature systems. However, the system complexity increases quite rapidly with efficiency, often resulting in considerable local maintenance requirements for the more efficient designs, as well as difficulties in technology transfer to user countries.

Photovoltaic Conversion

Photovoltaic methods are attractive because of their simplicity. Solar cell panels produce DC electricity which can be used directly for lighting and for powering buoys, TV sets, telecommunication equipment, etc., and with some additional circuits, for driving electric motors, pumps and the like. The panels are sometimes associated with wind-powered generators, and in many cases with storage batteries for continued electricity supply at night and during periods of rain, cloud cover, etc. Other advantages of photovoltaic conversion methods include: operation under limited cloud cover which does not interrupt electricity supply; no cooling circuits are needed in most cases; maintenance is very slight (Rodot and Vasseur, 1977).

On the other hand, solar cell fabrication involves rather complex processes, making technology transfer very difficult. The main obstacle to large-scale use of solar cells is, however, their high cost, which will be significantly reduced only by mass production. System costs are largely insensitive to economies of scale, because cost is basically proportional to the surface area of the solar panels and there is only a slight decrease in unit power costs as a system's size (power rating) is increased*. In addition, power requirements are difficult to satisfy when the electricity produced must be temporarily stored; batteries having only limited storage capacities.

The present-day cost of photovoltaic cells on the world market is about 7 to 10 dollars per peak-watt, according to the quantity ordered, with target maximum selling prices of 1 dollar per peak-watt by 1985 and 0.5 to 0.7 dollars by 1986, at which time world-wide annual production is expected to be several hundred peak-megawatts (Fig. 3.5). Such low costs will probably be achieved only by using fabrication methods and cell materials different from those of the current generation of circular, single-crystal silicon cells. Extensive public and private research and development in France and other European countries, in the United States and Japan, etc., are being carried out to investigate ways and means of improving cell efficiency. This includes continuous production methods involving pulling single-crystal silicon into thin ribbons; high-temperature deposition of thin layers of polycrystalline silicon on a substrate; use of amorphous silicon obtained by decomposition of silane (SiH 4), and of

Fig. 3.5 Photovoltaic Module Prices (After 14th Photovoltaic Conference, 1980).

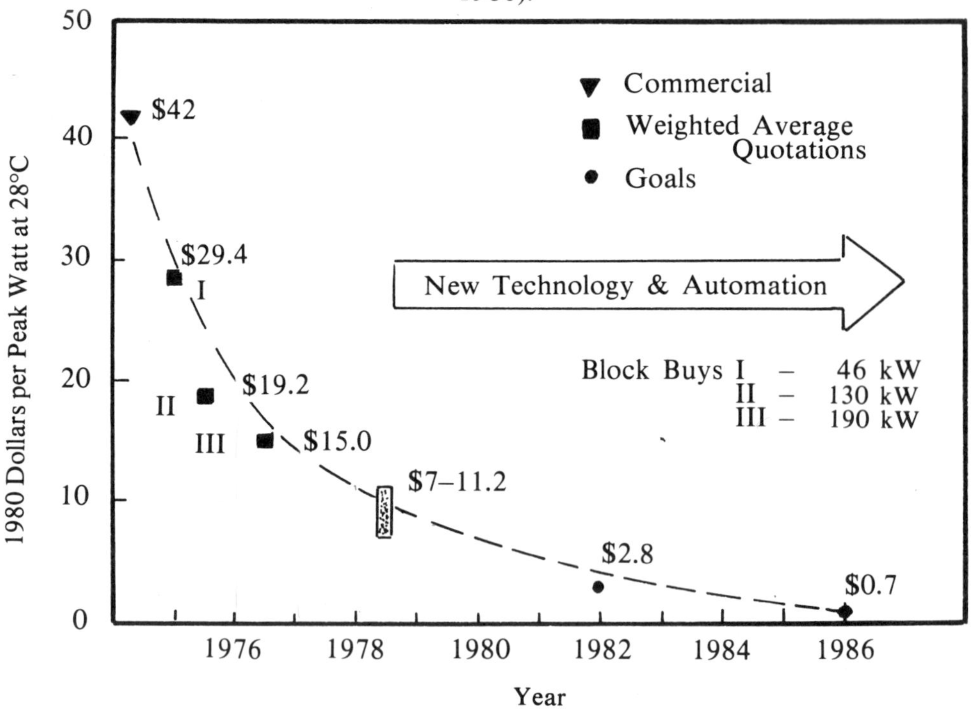

*Electricity costs are usually given in U.S. dollars per peak-watt, corresponding to the power available at the solar panel terminals with an incident radiation of 1 kW per square metre at 25°C. This should not be confused with the average power supplied to the shaft of a motor or alternator, for example, which is 4 to 10 times less on a 24-hour average basis, according to local climatic conditions (2 to 5 times less when considering only daily sunshine periods).

semicrystalline silicon; cells made of compounds such as gallium arsenide (GaAs), cadmium telleride (CdTe) and cadmium sulfide (CdS), and compounds of copper (Cu_2S, Cu_2Te and Cu_2Se); special cells with optical concentration of the solar radiation; so-called tandem cells, which associate several cells employing materials of differing spectral responses, to increase overall conversion efficiency. It is safe to predict that price reductions and technical improvements during the coming years will open up many new fields of application for photovoltaic cells. They are already suitable for small-capacity decentralized electricity production (from about 1 to a few tens of kilowatts). Solar cells factories have already been built in the United States with annual production capacities of several tens of megawatts.

Demonstration projects making use of photovoltaic conversion are under way in several countries. Examples are:

- a photovoltaic house at Valbonne close to Nice in France. This demonstration house has completely autonomous electricity production by a 5 kW photovoltaic generator for lighting, household equipment and appliances, and air conditioning. It is being built by the LEP-TRT group, with the assistance of the French National Space Research Centre (CNES) and National Scientific Research Centre (CNRS);
- an alpine refuge with a photovoltaic generator of the same capacity, to be built by the French company SERI-Renault;
- a 9 peak-kW photovoltaic system supplying electricity for lighting and medical equipment at the San hospital 450 km from Bamako in Mali, built with the help of the French Ministry of Economic Co-operation;
- development by the Martin Marietta company and the Solar Energy Research Institute of systems for supplying electricity to two villages in Saudi Arabia, using concentrating solar cells panels with a total power of 350 peak-kW.

Low-temperature Thermal Conversion

This is the conversion method developed by the French company Sofretes, and used in installations throughout the world, from the Philippines to Mexico (Girardier, 1973). In typical systems, flat-plate collectors with selective-surface absorbers raise the temperature of a primary heat-transfer fluid to about 80 to 95°C. This fluid is then used in a freon-based heat engine cycle to drive a motor or turbine. The heat source is the solar-heated primary fluid, and the heat sink is pumped groundwater or a cooling tower acting as an evaporative condenser. After numerous small-capacity (about 1 kW) applications in pumping water for livestock and human domestic use, and for irrigation, the system has been extended to decentralized electricity production, with outputs from roughly 10 to 100 kW.*

*Rather than indicating the capacity of a system by its instantaneous power output in watts, which varies according to sunshine conditions, the preferred method uses the daily power output in kWh per day. This daily output can be increased by associating the electricity-generating system with a heat storage facility, for continued power production in the absence of sunshine (storage, can involve the primary heat-transfer fluid or a secondary storage medium).

The low-temperature thermal technique is a proven method of power production. Well over 80 systems are operating or under construction in about 20 countries, with capacities from 5 to several hundred kWh/day. As sunlight is captured without concentration, use is made of both direct and diffuse radiation, and operation is possible with a certain amount of cloud cover (frequent in tropical regions, particularly close to the sea). Relatively simple techniques are involved, making technology transfer easy: for example, uncomplicated flat-plate collector designs exist using rugged, long-life materials capable of withstanding even the most severe weather conditions. Storing heat as hot water is a trouble-free method for construction of systems operating on a 24-hour basis. Furthermore, collectors can be used as the roofs of houses, shops and other buildings, significantly reducing total investment. Up to 1,000 kWh/day, the initial costs per kWh/day are amongst the lowest obtainable with alternative energy sources (except for very small systems). In addition, maintenance is minimal, being limited to periodic surveillance and inspection. The life span of a low-thermal system is a bit longer than for competing systems such as diesel generators. One of the first low-temperature thermal systems was installed in Senegal in 1962, and is still operating without trouble.

On the other hand, although low-temperature thermal systems are simple, equipment is more complex and cumbersome than with photovoltaic conversion, which is generally preferable for very low power applications. For efficient, competitive low-temperature solar thermal power generation, especially at high power ratings, it is essential that the solar system design be integrated and co-ordinated with other infrastructures and development aspects such as civil engineering, socio-economic development projects, etc. necessary for successful development of rural and isolated regions, by creation of new centres of activity, at which electricity production should satisfy the needs for development projects. For very large-power applications (over 1 megawatt), flat-plate collectors are excessively large, and system design and installation become very complex. High-temperature thermal conversion methods are almost always preferable in such cases.

The initial costs, usually expressed in U.S. dollars per kWh of daily output, vary according to the daily production capacity. Over the period 1974 to 1978, systems steadily increased in size, and initial costs were reduced by a factor of 4 to 5. They are now lower than for photovoltaic systems for capacities exceeding a few kWh/day. The costs are, however, slightly higher than the lowest costs for wind-powered generators under ideal wind speed and frequency conditions. There is plenty of room for additional technical improvement, which will lead to further price reductions. Other cost factors must be considered when comparing LT solar thermal systems with diesel generators, their main rivals for rural areas. The lifetime cost of a diesel generator is many times higher than its initial cost, because of unavoidable outlays for periodic engine maintenance and replacement, the high wages of skilled maintenance technicians, effects of inflation and political-economic factors on costs (labour, transport, fuel, etc.), and so on. For many isolated sites in Africa, Latin America, etc., solar systems have proven more cost-effective than diesel generators, despite their much higher initial price, because they have virtually no such additional life-long costs.

Several low-temperature solar thermal conversion systems have been installed in some countries. For example, a 30 kW solar power station was installed at San Luis de la Paz, Mexico and has been operating since 1975. The station has 1,500 m^2 of collectors and uses a Freon II heat engine cycle. It is used for pumping 900 m^3 of water per day for domestic purposes (for a community of 15,000 persons), and for irrigating 30 hectares of experimental farms. The station is also used to provide electricity for lighting. Another example is the Dire facility in Mali commissioned in 1979. The facility consists of a 70 kW solar power station on the Niger river, built under the "Sahel Alternative Energy Sources" programme. It has 3,000 m^2 of collectors, two 1,000 m^3/h irrigation pumps, and 15 kW turbogenerator. The station is used for pumping 9,000 m^3/day of water for irrigation of 150 hectares, plus 600 m^3 per day for domestic use by an agricultural-tourist complex; refrigeration and space cooling for a small hotel sheltered under the collectors (30 million negative calories per day) are planned.

Low-temperature solar thermal systems for decentralized production of electricity have also been installed in other countries. For example, a solar plant was built in Bakel, Senegal by the Government of Senegal, with American and French assistance to power an electric pump for withdrawing 8,000 m^3/day of water from the Senegal river for irrigation of 100 to 150 hectares. A low-temperature solar thermal power station with heat storage, operating on a 24-hour basis (capacity: 250 kWh/day, 35 peak-kW) was built in Riyadh, Saudi Arabia by Sofretes. The station serves as a demonstration plant for electric power production and water pumping at an agricultural research centre.

High-temperature Thermal Conversion

In a high-temperature solar thermal conversion system, the temperature is considerably higher than in the LT system described above, thus shifting operating temperatures closer to those of conventional (non-solar) thermal power units. Several techniques of varying sophistication are available for attaining such high temperatures. All are based on concentrating the received solar radiation. The systems are characterized by the radiation concentration factor, the temperatures of the fluids used in the boiler, in the heat engine and for heat storage, and by the overall thermodynamic efficiency.

High-temperature solar conversion systems can be divided into two broad categories:

- modular, *distributed receiver systems*, using parabolic or cylindrical-parabolic collectors,
- non-modular, *central receiver systems* (also known as *solar tower systems*), in which a large number of heliostats concentrate the solar radiation at a central boiler.

In both categories, the components (collectors, heliostats or boiler) must be dynamically oriented towards the sun by rotation around one or two axes: i.e., they must "follow" the sun during its daily and seasonal movement in the sky.

Because of the higher temperature of the heat source, high-temperature solar conversion systems are thermodynamically more efficient than LT thermal systems. Very high capacity plants can be built (up to several tens of megawatts), with turbogenerator operating conditions close to those in conventional power plants. However, limitations on the maximum capacity to be achieved are imposed by the thermal characteristics of construction material and the accuracy in sun tracking, which influences the size of the receiver field. For example, the 1 MW E.E.C. power station in Italy is now in operation, and the 2 MW Themis solar tower plant sponsored by the French National Scientific Research Centre (CNRS) will have a field of 360 mirrors (heliostats), each 45 m^2 in size, and covering 5 hectares (Etievant and Pharabod, 1977).

On the other hand, since efficient concentration of solar radiation is possible only when the sky is absolutely clear, HT thermal methods are difficult to use in regions where clouds and haze are frequent. In such cases, it is necessary to have large heat storage systems to ensure uninterrupted electricity production during unfavourable climatic conditions. The high accuracy necessary in sun tracking requires very rugged and fairly complex receiver steering systems, and this makes technology transfer difficult. However, distributed receiver techniques have been specially designed to facilitate technology transfer as is the case with, linear focussing systems with stationary collectors and mobile boiler, for example. A 100 kWe unit under construction in Corsica will have a field of these collectors occupying about 4,000 m^2 of land with 1,200 m^2 of mirrors.

Several types of HT thermal conversion unit are under construction in France, other European countries and elsewhere, mainly for demonstration and experimental purposes, with true commercial use planned for a few years from now. These include many designs of distributed receiver system, as well as the following solar tower units: International Energy Agency project for a 500 kW sodium-cycle unit in Spain; two 1 MWe units in Japan; the French 2 MW Themis project mentioned above; the 5 MWt Central Receiver Test Facility at Albuquerque in New Mexico; and the Central Power Plant at Barstow in California (10 MWe).

Solar-generated electricity was first supplied to a national power grid from the Odeillo unit in France, in November 1976. Since then, much progress has been made, and long-term prospects include electricity costs of about 20 cents per KWh under good direct sunshine conditions, following series production of components. This cost level will permit use not only in developing countries, but also in the sunniest regions of industrialized nations (Southern Europe, Australia, California, Arizona, New Mexico, etc. in the United States, and so on), as solar electricity will then be cheaper than that generated with fossil fuels (coal and oil), though still more expensive than nuclear electricity (some countries may, however, be prepared to accept higher-than-nuclear costs for political and other reasons).

Applications of Terrestrial Solar Electricity Systems

All the methods described above are basically designed to supply electricity to remote areas with favourable climatic conditions; the best system to be selected depends on the requirements, the size of the community to be served, etc.

For very small villages and hamlets with a few tens to a hundred or so inhabitants, it is preferable to introduce simple solar techniques adapted to individual needs, rather than the more costly and complex systems better suited to collective facilities such as large schools and hospitals. In such cases, very small scale production of electricity, by a photovoltaic system or a low-capacity LT thermal system is appropriate.

On the other hand, for larger communities of several hundred to a few thousand persons, requirements for water, electricity, cooling/refrigeration, motive power, etc. necessitate generating up to several thousand kWh/day of power (experience has shown that demand tends to increase rapidly as soon as electricity becomes available). Low-temperature solar thermal systems with flat-plate collectors and (if direct sunshine is abundant) medium-sized high-temperature thermal systems which can also supply process steam for small industries, are most suitable in such cases.

For towns with tens of thousands of inhabitants, solar energy is in direct competition with conventional energy sources; the current costs of solar-generated electricity are, however, too high. Extensive research and development work underway, on both LT and HT thermal methods, will probably lead to cost-effective solutions within a few years.

Thus, in the near future, solar energy will play a growing role in satisfying parts of the electricity needs of small and medium sized communities. Even so, solar electricity will represent only a very modest contribution to the world's total energy consumption. The benefits of solar energy will become apparent in the developing countries where geographical and climatic conditions are generally most favourable. It is, however, possible to imagine a situation in which intensive research and development and systematic use of all available collection surfaces (roofs of houses and other buildings, uncultivated areas in the countryside, etc.) will enable even a European country like France to profit from widespread, extremely decentralized solar electricity production.

Solar Power Satellites

Another possible future development is the use of Solar Power Satellites, which are the subject of a great deal of research in the United States and some other countries. These satellites would be placed in a geostationary orbit; being equipped with enormous solar cell arrays (many tens of square kilometres), they will beam the collected solar energy to earth in the form of microwave or laser beams for reception by large antennas, and then conversion to electricity (U.S.DOE, 1978). The feasibility, lead times, fabrication, launch and operating costs for such satellite systems have been carefully evaluated (Glaser, 1979); the earliest possible date for a full-scale commercial use has been given as 1996.

Besides the technical and financial difficulties encountered in such a programme, there are several environmental impacts to be anticipated: use of land, water, oxygen and hydrogen resources and other raw materials for the satellites, launch vehicles and ground facilities; noise pollution due to the large number of launches necessary (because of their vast size, the satellites would be sent up in "bits and pieces" for in-space assembly); air pollution and effects of microwave beams and rocket-emitted products on the ionosphere and stratosphere; etc. Nonetheless, there seems to be no real technical or scientific barrier to construction of power satellites, and they could represent a promising future solution to the challenge of large-scale electricity production, from renewable sources of energy.

WATER PUMPING AND IRRIGATION

The main objectives of rural development in the Third World is to satisfy the basic needs of the population of such areas, improve their living conditions and achieve at least a minimum of economic growth by means of agriculture (crops and livestock), traditional crafts, and small industries. One of the main obstacles to such development is often the difficulty of procuring and distributing water of adequate quality and quantity for domestic use, livestock, and for crop irrigation.

The lack of water is generally most critical in the very sunny areas of the earth. Such areas suffer from periodic droughts, from the lack of adequate transportation systems and from the lack of qualified labour for maintenance and operation of equipment at isolated sites. These regions, however, offer the best conditions for the application of solar techniques for water desalination (discussed later), and for water pumping and irrigation.

For water pumping and irrigation, a range of solar techniques are available, depending on the required power. These include thermal conversion methods, with and without solar radiation concentration, photovoltaic conversion as well as biogass-run motors and wind-powered generators. In all cases, the energy-producing system must be coupled with a suitable pump or pumps.

It should be noted that the conversion of solar energy to mechanical power for such applications was investigated by many 19th century scientists in France, India, United States and elsewhere. However, it was not until the 1960s that industrial systems have been successfully developed and marketed; among the pioneers in this field is the French company Sofretes, which has installed solar-powered pumping units in many countries in Africa, Latin America and the Middle East (Clemot, 1978). Thermal conversion systems have also been installed in the United States (Arizona and New Mexico) since 1975, with the support of the Energy Research and Development Administration (ERDA). With the exception of many photovoltaic units about 1 kW capacity and a 25 kW project using 120,000 solar cells in Nebraska and a French project of the same capacity at Montpellier, all the larger systems installed to date have employed thermal conversion with capacities ranging up to 100 kW or so. Examples of these are the Dire installation in Mali (Pheline, 1979) and the Sandia Corporation 200 HP project in Arizona.

Solar water pumping and irrigation could become more common by the adequate transfer of technology. The heavy and expensive components of the system should be manufactured in the recipient countries. The systems must not be too complex nor involve overly sophisticated technologies and manufacturing processes. Technology transfer is undoubtedly easiest with flat-plate collector systems, including those with selective surfaces for enhanced efficiency. Concentrating collectors providing a small to medium amount of solar radiation concentration can also be used, to increase operating temperatures, and hence efficiencies. One of the most important factors in comparing solar techniques with more conventional power supply methods (e.g., diesel engines) is the distance and mode of transport (and risks) between the site and the most accessible fuel depot. In Africa, the solar option is favourable as soon as this distance exceeds about 100 km, at current costs. This distance can only decrease in the future, because solar methods have the inherent advantages of "leave-alone" operation (virtually zero maintenance, no requirement for skilled labour, etc.) and independence of their lifetime costs from increases in the price of petroleum-based products.

GREENHOUSES, PRODUCT DRYING, AND COOKING

Any surface is heated when exposed to sunlight, because light is absorbed and transformed into heat. As explained earlier, solar thermal collectors are basically devices designed to optimize this conversion of solar radiation into thermal energy and transfer the collected heat to a suitable fluid (air, water, etc.) made to flow over the heated surface. The temperature at which heat can be extracted from a solar system depends on the type of the collector. With ordinary flat-plate collectors, heat can be obtained at temperatures up to about 150°C. For higher temperatures, it is necessary to use selective surfaces or to concentrate the sunlight before collection. Linear focussing collectors of the parabolic trough and Fresnel lens types give temperatures up to around 300°C, while point focussing collectors can reach very high temperatures (600°C or more).

The potential applications of solar heat in agriculture and in industry are numerous within the above-mentioned temperature ranges. These have been evaluated in depth by many authors and organizations (ERDA, 1977; Mitre Corporation, 1979) and have been the subject of several international conferences (EEC, 1979; Congrès solaire international, 1979). For some applications, the thermal energy is converted to mechanical energy or electricity, or used as the heat source for cooling or refrigeration, whereas in other cases the heat is directly used. Solar space heating and solar water heating have already been discussed. The present section concerns three other applications of solar heat: crop growing (greenhouses), product drying, and cooking.

Solar Greenhouses:

The use of greenhouses is common in countries with temperate climates. They allow some crops to grow in winter without the risk of damage by frost and freezing temperatures; for this purpose greenhouses should, however, be equipped with an internal heating system. Greenhouses are also used in very sunny countries to obtain a cooling effect. Evaporative type cooling systems are usually used, but they have the inconvenience of consuming a great deal of water. A new process developed in France (Balligand *et al.,* 1979) limits the water consumption and provides better control of the greenhouse temperature. In this process (Fig. 3.6), the walls of the greenhouse are made of double-glazed panels, with special filtering glass used for the outer glazing, and ordinary glass for the inner glazing. The filtering glass stops the infrared component of solar radiation, while allowing the visible and ultraviolet light necessary for plant growth to pass through. Heat accumulated in the walls by infrared radiation absorption is removed by water circulated through the space between the filtering and non-filtering glass. The heated water is stored in a basin, and later recirculated back through the glass panels or in plastic ducts placed on the ground, when the greenhouse must be heated (e.g., at night). In addition to reducing the heating costs, this technique has the advantage of transferring excess heat to the storage basin during sunny periods, thus helping to prevent excessive plant temperatures, and reducing water consumption and ventilation costs.

Fig. 3.6 Solar Hothouse Using Filtering Glass. On the Left: Hot Water Recirculation is through the Walls. On the Right: Hot Water Recirculation is through Plastic Ducts Laid on the Ground.

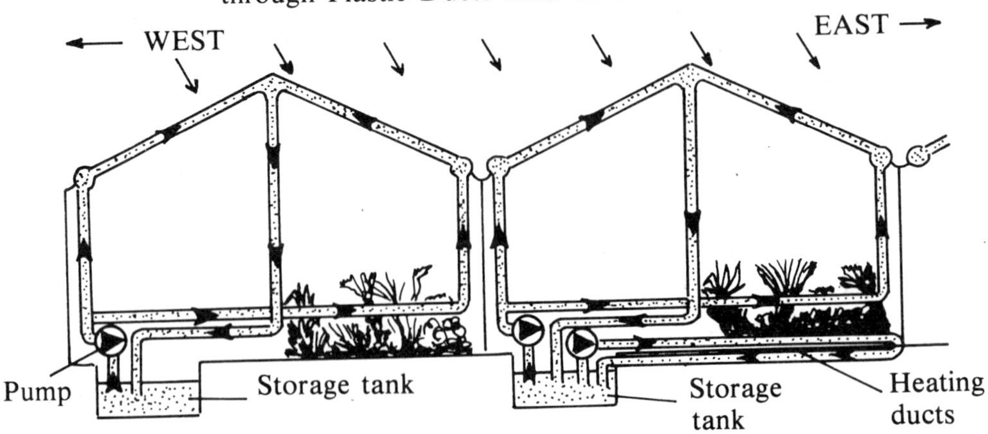

Solar Drying

Of all the direct uses of solar energy, sun drying of crops is perhaps the most ancient and widespread. The customary technique involves spreading the material to be dried in a thin layer on the ground, and exposing it to the sun and wind. Copra, grain, hay, fruits and vegetables, etc. are still dried in this manner in many parts of the world, including the industrialized countries. An innovation adopted in recent years for fruit drying involves placing the fruit in specially designed racks for controlled exposure to solar radiation and wind, and improved handling, resulting in better process control and product quality. The term solar drying has also come to mean the process whereby agricultural products are dried not by direct exposure to the sun, but by solar-heated air in more protected surroundings. This process is of special interest for soft fruits, which are particularly vulnerable to attack by insects, because their sugar content increases during drying.

The overriding reason for the universal use of sun drying is that it is the simplest way of drying crops such as fruits and vegetables in regions with abundant sunshine and where the post-harvest season is characterized by low relative humidity and little or no rainfall. The last twenty years or so have seen several attempts to apply modern knowledge and technology to improve traditional methods to extend the application of solar drying. Although this had not led to any significant commercial manufacturing of solar drop dryers, it has produced a number of experimental designs which are being used on a small-scale (UNIDO, 1978). These range from the use of solar-heated air in more or less conventional air dryers (various designs have been developed in the United States, Turkey, Pakistan, Thailand, Canada, Brazil, Argentina, etc.) to methods designed and successfully tried out in India (Pattanayak *et al.,* 1978; Muthuveerapan, 1979) and Trinidad, which combine direct sun drying and air drying by placing the products in flat-plate type collector-dryers. Fig. 3.7 illustrates an air dryer developed in Thailand for drying rice.

Future developments in solar drying will include improved combination of collector-dryers and flat-plate type air heaters, and their association with heat storage facilities. Present research on the design and control of solar drying processes for particular crops and materials will probably lead to new applications and improved processing of food resources in the developing countries (NAS, 1976; Forster and Peart, 1976).

Solar Cooking

Many types of solar cookers have been developed since the beginning of the present century (Fig. 3.8). In general, they may be divided into two categories:

- concentrating parabolic and spherical trough or dish collectors in which the heat concentrated at the collector's focal point directly heats up either a vessel containing the food, or the food itself,

– ovens and food warmers, which are basically insulated boxes with transparent covers in which the solar energy is collected directly or after reflection from a special surface.

Fig. 3.7 Experimental Solar Rice Dryer, Pathumthani Province, Thailand.

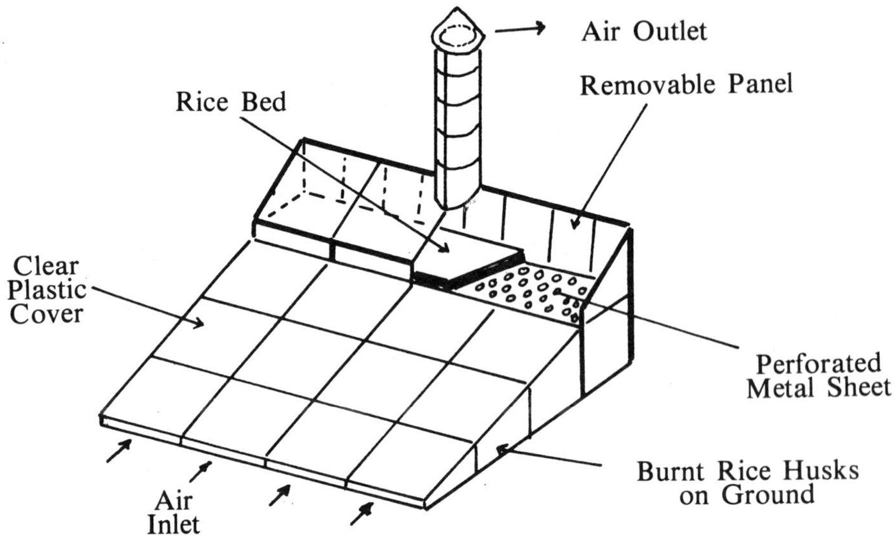

A parabolic mirror type solar cooker was developed and marketed in India as early as 1950. Although it was sold for a low price, it did not attract much interest, mainly because few housewives could accept the idea of a new cooking process different from the ones employed by generations of ancestors. Other negative factors were the variation in efficiency (temperatures and cooking times) according to climatic conditions, and the need to use other cooking methods in the absence of sunshine (rainy days, night time, etc.). Based on several published evaluations of solar cookers (Walton *et al.*, 1978; Hoda, 1978; Channabasappa, 1979), the main requirements for their success in developing countries may be summarized as follows:

(*a*) solar cookers must provide energy at a sufficient rate and temperature to properly cook desired types and quantities of food,

(*b*) the cookers must be sturdy enough to withstand rough handling, wind, and other hazards,

(*c*) they must be sociologically acceptable and fit in with the cooking and eating habits of the people, for example, to allow cooking in the shade and (preferably) at times when the sun is not shining,

(*d*) solar cookers should be easy to manufacture using local materials and labour, and

(*e*) the price of solar cookers must be lower than or competitive with other cooking devices.

At least two advanced solar cooker designs have been proposed to overcome the difficulty that solar cooking is normally possible only when the sun is shining and when the cooker is directly exposed to the sunshine. The first design uses a heat-transfer system to allow cooking inside houses or in shaded spots, whereas the second makes use of a heat storage system to permit cooking when the sun is not shining, or at night. Both these types of cooker are considerably more expensive than simpler open-air units, but they are also more acceptable to users. It is believed that large-scale use of solar cookers (if they become acceptable) in some developing countries will markedly reduce the dependance on firewood and, hence, the desertification process.

Fig. 3.8 Different Types of Solar Cookers.

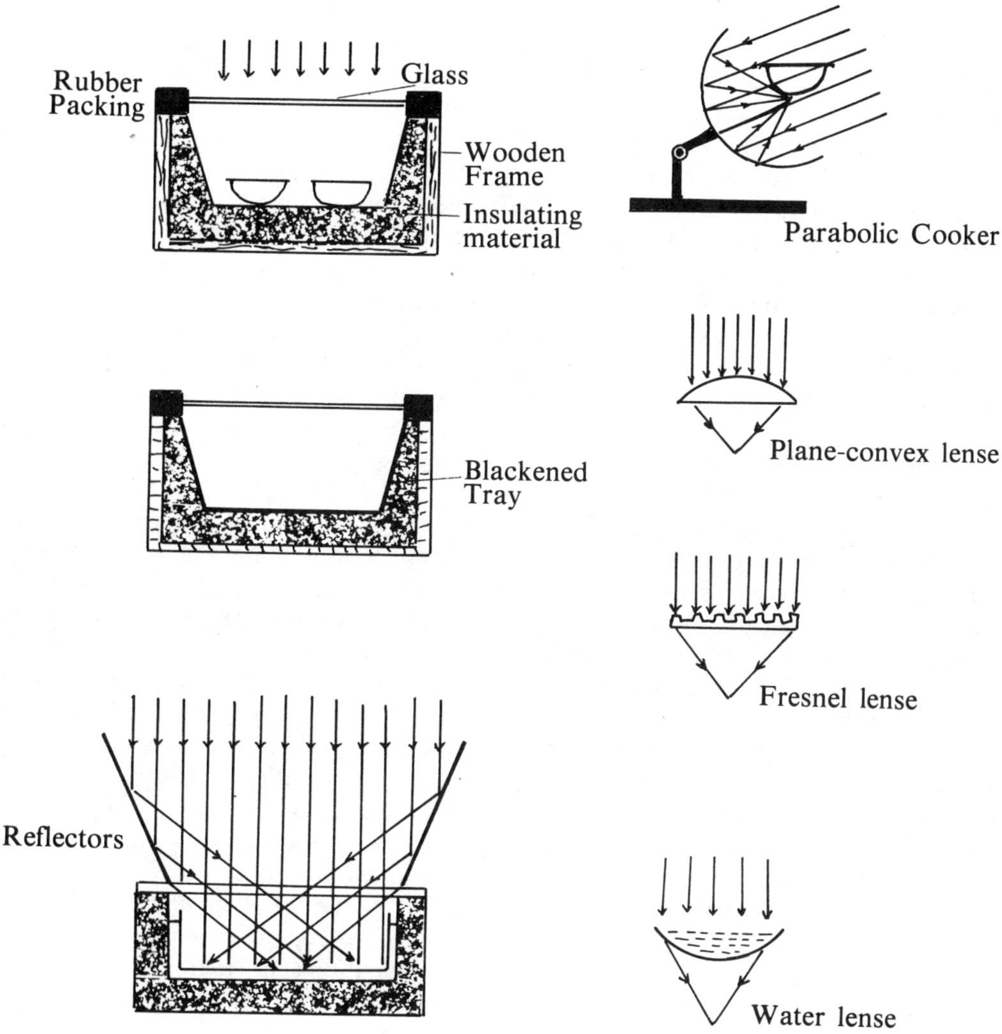

DISTILLATION AND DESALINATION

The World Health Organization (WHO) specifies a maximum salinity of 500 mg/litre in water intended for domestic use; for most industrial and agricultural uses, the water salinity must not exceed 1,000 to 1,500 mg/l. Natural water resources satisfying these requirements are very limited, especially in arid and semi-arid zones. For this reason, considerable efforts are being made to develop cost-effective methods of water distillation and desalination, particularly those making use of solar energy.

The century-old process of the basin still has been modified and adapted to modern materials. It consists of a shallow pool of brine from which water is slowly evaporated and condensed on the underside of cooler glass covers; the water runs to troughs at the edge of the basin, and from there to storage. The concentrated brine is then run to waste as new saline water is added to the basin. The idea was first applied in 1892 at Las Salinas in Chile, in a plant supplying drinking water for animals working in nitrate mining and transport– the plant operated for about 30 years (NAS, 1976). This and other modern developments in solar distillation are being used on a small scale in many parts of the world to supply water to small towns and isolated motels, for example, in Australia, in the Mediterranean, Caribbean and Middle-East regions (Channabasappa, 1979).

Current research and development aims at selecting construction material and perfecting designs to decrease initial costs, in addition to limiting maintenance and increasing the life span of the plants so as to reduce the total cost of distilled water. At present, solar distillation requires a relatively large capital investment per unit of capacity, but operating and maintenance costs are very low in properly designed and built systems. The water costs depend primarily on the still productivity, its life span, initial investment, and amortization and interest rates. Still productivity is dependent on the intensity and duration of sunshine, but experience shows that a typical still yields about 4–5 litres/m^2 per day, with some variations according to climate and still design. The typical service life of a still made of concrete, glass and other durable materials is at least 20 years. Other still designs use less durable material that must be replaced or renewed from time to time. Although the influence of productivity, service life and maintenance cannot be ignored, the present high cost of distilled water production is mainly due to the very high initial capital investment involved, and there seems little prospect that large solar stills will be competitive with large desalination plants using conventional energy sources, unless fuel prices escalate far beyond today's already high levels. However, in situations where a community or industry requires only small amounts of water (say less than 20,000 litres per day), the solar still can be more economical than a conventional desalination unit. This is particularly true (*a*) in small communities where potable water is unobtainable by any means except at very high costs, (*b*) for some industrial materials-processing applications in regions where all available water is brackish, and (*c*), for livestock watering in areas where grazing is possible only if fresh water is supplied. With a few exceptions, all existing solar stills were built for these applications.

An interesting recent development is a solar desalination process based on a design of a greenhouse, and is intended for arid desert zones (Balligand *et al.*, 1979). In addition to producing fresh water from sea or brackish water, the process helps keep the greenhouse cool during the day, and warms it at night. The fresh water is used for plant watering, with any surplus to be used for other purposes. Fig. 3.9 shows such a greenhouse. The key element in its design is the use of a special filtering glazing, which stops the infra-red component of sunlight, allowing the visible and ultraviolet light to pass through for plant growth (photosynthesis). During the day, saline water trickles over these special glass panels, taking up the heat generated by absorption of the IR radiation; the water is then routed to a hot water storage pit. During the night, when it gets quite cold, this stored hot water is recirculated to heat the greenhouse. While trickling over the glass panels, some of the saline water evaporates: the resulting vapour condenses as fresh water on the inside of the greenhouse's outer wall, from which it is recovered for plant watering.

Fig. 3.9 Desalinating Solar Greenhouse.

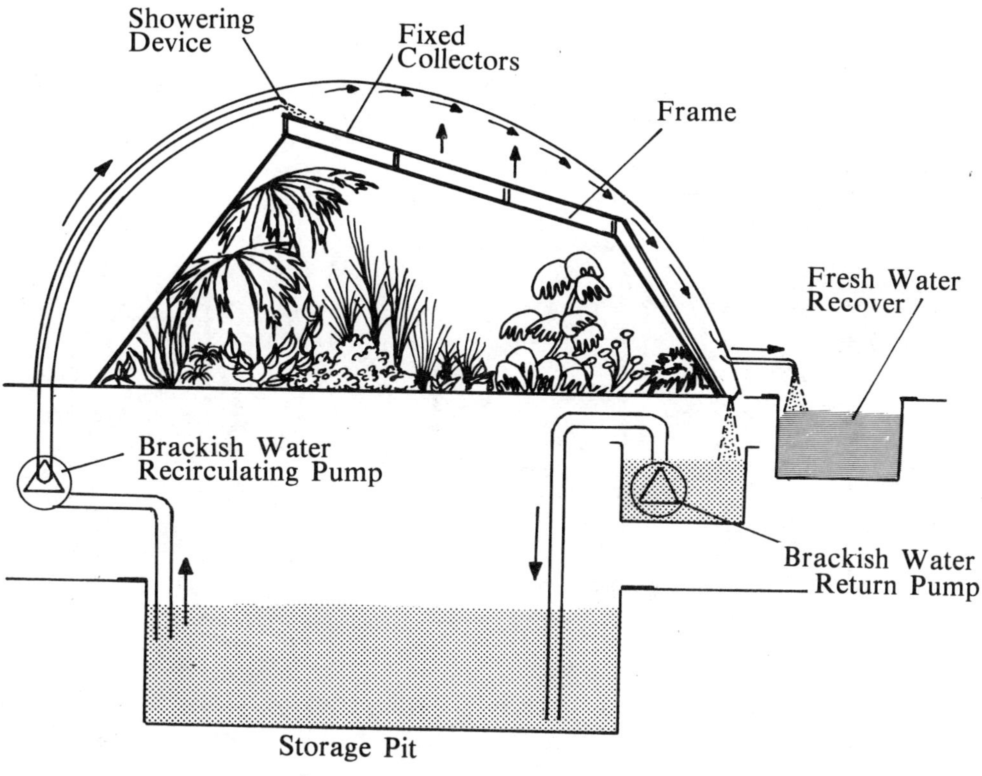

Some recent installations have also used solar radiation as the energy source for conventional desalination processes (Maurel, 1979). As solar energy is still rather expensive, the desalination method to be selected must consume small amounts of energy. This can be achieved either by using very high efficiency processes, or ones using membranes (e.g., electrodialysis or reverse osmosis). The best prospects lie in the joint use of solar-powered pumps with reverse osmosis or electrodialysis. For brackish water desalination, these processes consume small amounts of energy (2 to 3 kW/m^3 for water with 2.5 g/litre salinity); this consumption can be further reduced by increasing the membrane area.

The advantages of reverse osmosis over electrodialysis lie in the fact that it makes direct use of the mechanical energy produced by the solar pump, saving, therefore, the cost of an alternator (see Fig. 3.10). Moreover, extremely small amounts of energy are consumed in the process due to the following:

- it is a very simple process, basically involving a single HP pump, with hardly any auxiliary energy-consuming equipment,
- the current standard operating pressures (about 30 bar for brackish water) could be reduced by a factor of 2 or more; the reduction in membrane selectivity with reduced pressure being compensated for by decreasing the conversion rate, and
- the hydraulic energy of the HP water pump at the output of the unit could be recovered in a Pelton turbine, for example, instead of being wasted through a pressure-reducing valve.

Fig. 3.10 Solar Desalination Plant combining The Reverse Osmosis Process and a Solar-powered Pump.

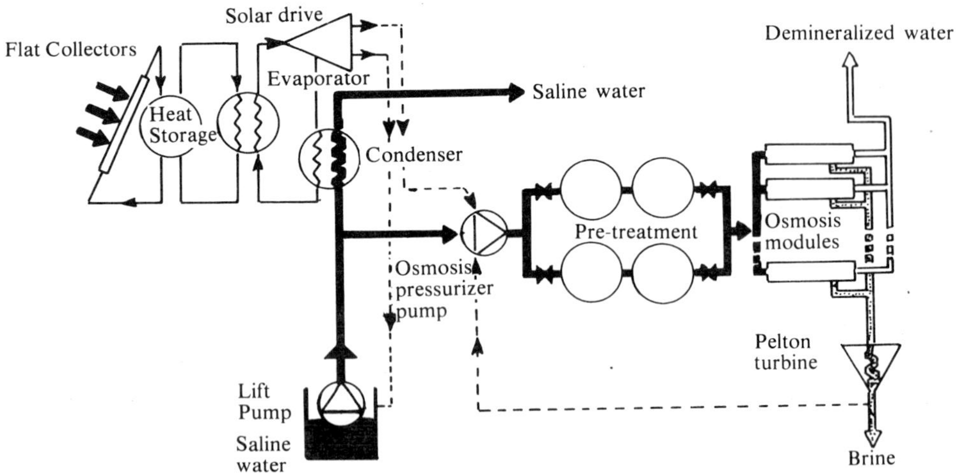

For brackish water with 2.5 g/litre salinity, these improvements lead to an energy consumption close to 1 kWh/m^3 (compared to about 3 kWh/m^3 in conventional units). Such an improved desalination unit has been installed, close to a phosphate mine on the Red Sea in Egypt. The reverse osmosis unit is coupled with a 10 kW solar-powered motor; the output is 60 m^3 of fresh water per day from 2 g/litre brackish water.

Another possibility is to combine a multistage flash evaporation distillation process with a solar heat-producing facility using concentrating cylindrical-parabolic collectors. This system was selected for a demonstration plant in Mexico, built with the assistance of the Federal Republic of Germany (Fig. 3.11) (Manjarrez and Galvan, 1979).

Fig. 3.11 Solar-powered Multistage Flash Evaporation Desalination system.

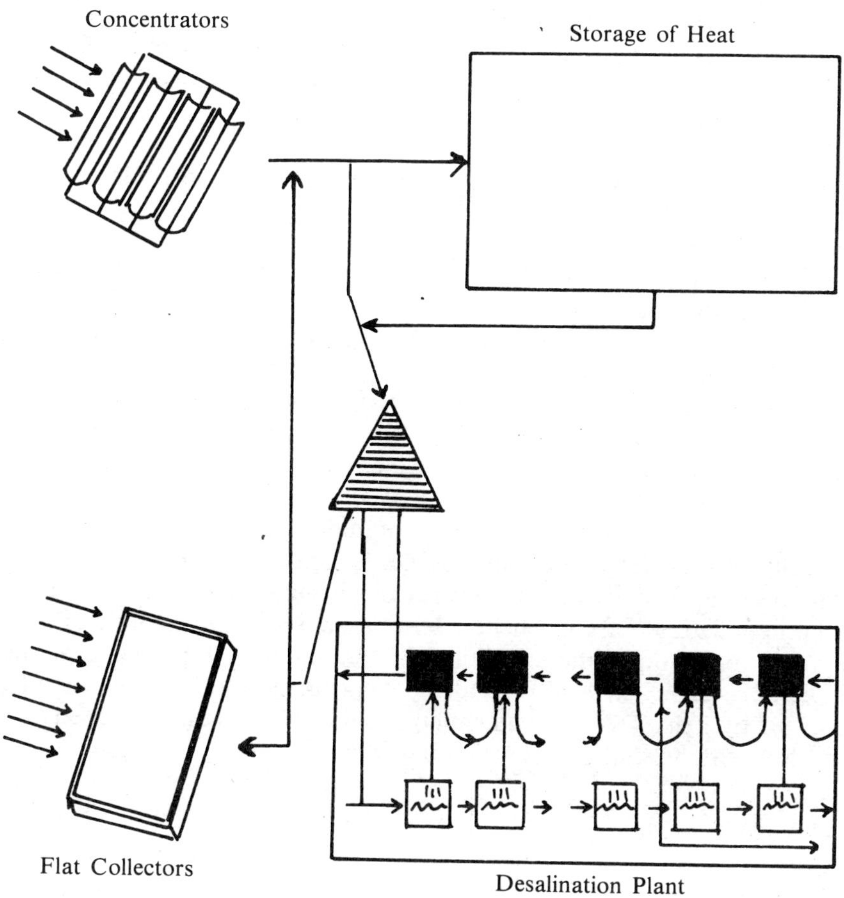

If the pilot plants described above will give satisfactory results, with respect to costs and reliability, it is anticipated that higher-capacity units will be built worldwide during the present decade.

INDUSTRIAL PROCESS HEAT

The solar techniques described in the preceding sections can heat air, water, organic fluids, etc. to temperatures from about 30 °C to hundreds of degrees. A small number of pilot plants has been built to demonstrate the use of this heat in industry. However, cost competitiveness with more traditional heating processes has not yet been demonstrated with certainty. Another handicap is that the solar system must almost always be associated with a backup conventional heating system, because it is rarely acceptable to interrupt industrial production or processing during the absence of sunshine.

Despite these difficulties, continuing increases in the prices of petroleum products may well make solar energy at least marginally competitive in the near future for numerous applications. In this respect, it should be noted that in industrialized regions such as Europe, industrial use of heat at temperatures within the range of those easily obtainable with solar systems (i.e., below about 250 °C) represents roughly a quarter of the total energy consumed by industry (Durand, 1979).

The following are examples of ongoing experiments to use solar process heat by industry:

- pasteurization of milk (France);
- use of solar-generated steam in brewing and food processing (Australia);
- tobacco drying (Argentina);
- solvent evaporation based extraction of oils from tropical plants (Indonesia);
- large-capacity refrigeration using the ammonia-water absorption cycle (Spain);
- distillation of marcs and lees of wine, using 100 parabolic concentrating collectors (EEC project in France);
- textile drying with process steam obtained with parabolic concentrating collectors (US Department of Energy Project in Minnesota).

Although far from complete, the previous list illustrates the wide variation of industrial applications of solar heat. Mention should also be made of very high temperature applications (metal melting, fabrication of high-purity ceramics, etc.) developed during the 1970s at the Odeillo solar furnace in France by the team of scientists led by Felix Trombe, and which will probably one day be used on a very large scale (Trombe and Le Phat Vinh, 1973).

ENVIRONMENTAL IMPACTS OF SOLAR ENERGY SYSTEMS

Domestic Applications

The use of solar techniques in homes and buildings for space and water heating, space cooling and air conditioning has no real, negative environmental impacts specially in rural areas, provided the solar systems are correctly designed and installed, respecting building codes, elementary health and safety precautions, etc. For example, in a solar water heating system, the primary heat transfer circuit containing an organic anti-freeze compound must be separated from the circuit which actually supplies hot water to users. A number of legal and regulatory measures have already been formulated in this respect (AFEDES, 1980), and others are under preparation. When correctly used, solar energy then helps to save non-renewable sources of energy and reduce pollution due to the combustion of fossil fuels.

In urban areas, however, the use of solar energy systems may be limited by problems of interaction and graceful integration with the existing urban environment. In space heating and cooling, for example, orienting new buildings to optimize the collection of available sunlight can create problems. It is rarely possible in existing urban areas (and even in new communities), the placement and orientation of buildings being dictated more by local topography, existing main roads, and the location of sewers, water supply systems and power lines, etc. than by the position of the sun. Proper orienting of buildings for solar energy utilization would probably increase their costs (longer driveways, less efficient land use, longer electricity and telephone lines, etc.) (EPA, 1976; OTA, 1978). Also, a building constructed to take maximum advantage of the sun would not always have enough roof or wall area available for the solar collectors, in which case additional collectors would have to be placed in arrays on the ground close to the building. Of course, it is possible to locate the collectors at some distance from the building(s) served, but this reduces the possibilities of multiple use, and increases the costs of transporting the solar energy to its points of use. Of course, designing a new residential area or commercial centre for maximum use of solar energy is easier than adopting solar techniques for an isolated building within an existing, dense urban community. Buildings can be oriented properly, height restrictions can be used to minimize shading, and areas set aside for collector fields can be chosen for minimum disruptive effects on the community.

Solar energy utilization will certainly influence residential and commercial architectural styles (design of collectors as integral parts of roofs, use of concentrating surfaces, and so on). These effects will, however, probably be mitigated by the fact that public tastes are conservative and not very susceptible to rapid change, especially in the case of single-family homes, and because building codes tend to discourage innovation. Solar features are likely to be incorporated first into accepted, existing building styles, with a progressive (but probably slow) trend towards new types of homes and buildings taking full advantage of solar energy for space and water heating, and for space cooling.

Direct thermal pollution of the environment due to the use of solar energy in residential and commercial space heating, water heating and space cooling will be minimal. Space heating and hot water systems are closed systems which do not reject heat to the exterior. Solar cooling systems do involve heat release and limited cooling water discharge, but these are not expected to have any significant impact on the environment, (many cities have already become "heat islands", partly because of the releasing of heat that accompanies many forms of human activity).

However, one area that requires further study is the possible impact on weather and air pollution of large areas of solar collectors, which would capture some of the sunlight normally reflected from buildings back into the atmosphere. This might affect temperature gradients in the air, which are significant determinants of wind, cloud cover, atmospheric mixing and (ultimately) air quality. There could even be a feedback effect whereby increased cloud cover reduces the effectiveness of the solar collectors (EPA, 1976). The environmental impacts of large storage systems must also be carefully studied. Such systems are probably the key to future widespread use of solar heating for collective purposes; and some may have potential environmental impacts (for example, the possible alteration of the soil's microbiological equilibrium by large underground heat stores).

ELECTRICITY PRODUCTION

Solar thermal power plants do not generate gaseous, liquid and solid wastes like fossil fuel and nuclear units, and they are relatively "neutral" insofar as excess heat release is concerned: i.e., the heat left at the site is similar to what would have been deposited by solar radiation if no plant were there. For example, a solar tower plant creates an excess of about 0·25 MWt of heat per 1 MWe of electricity, which is very low in comparison to the figures of 2·1 MWt for a light water reactor and 1·7 MWt for a fossil fuel plant (Caputo, 1977). The heliostat field of such a solar tower facility could produce local changes in the net albedo, energy balance, moisture balance, low-level wind patterns, and air and surface temperatures. Although the exact impact of such changes on the microclimate at and immediately surrounding the solar power plant has not yet been completely assessed, rough calculations show that for a plant installed in a desert, the climate cannot be changed significantly (DOE, 1977; Collection Environment, 1974). From the solar radiation falling on a desert, 40% is reflected (albedo) and 60% is absorbed by the lower atmosphere and the ground. The presence of an absorbing surface would reduce the reflected part to say 20% and if the 80% absorbed fraction were transformed in an adjacent plant, with a conversion efficiency of 25%, then 60% would be returned to the environment ($0{\cdot}8 \times 0{\cdot}75 = 0{\cdot}6$), and 20% used in generating electricity. In other words, the total energy input to the environment remains the same, and all the energy used by the plant is drawn from the reflected part. The albedo is, however, modified, but without harm to the environment, because the surface areas involved are small in comparison to the total surface of the earth. It has been pointed out that the total energy requirements of the United States at the beginning of the 21st century could be supplied by converting just 2% of the country's farms to solar farms, with 12,000 km^2 of collectors spread over 24,000 km^2, assuming an efficiency of 20 to 25%, and suitable heat storage facilities. This surface area is only 1/400th of the surface area of the U.S.A. If the same proportion applies to the entire globle, then the world's total energy needs could be supplied at the cost of a reduction in the earth's albedo of just 0·03%:

$$\frac{1}{1 - 0.5.1/400.0.25} \times 100 = 0.03$$

The factor of 0·25 is the proportion of emerged land in the total surface of the earth, the rest being oceans and seas. This new albedo would result in a temperature rise of 8/100 °C, as compared to an increase many times greater if fossil fuels were burned instead of using solar energy.

Some problems may arise from the use of toxic organic fluids in solar power plants for heat transfer and storage, and special precautions will have to be taken; such problems are, however, similar to those already existing in the chemical industry.

A special difficulty with mirror (heliostat) systems is the need for operating rules and safety measures to prevent accidental convergence of several light beams to a point other than the central boiler, as this could ignite fires or temporarily blind airplane pilots. It will be necessary to issue and strictly enforce special limitations on overflying these types of solar power stations. Although work is continuing to precisely evaluate potential damage to the eye by such concentrated light beams, first results suggest that it is not a very serious problem, providing elementary precautions are taken (Young, 1979).

The possible potential environmental impacts of solar power satellite systems should be studied in detail. Particular attention should be given to impacts such as:

- effects of microwave beam propagation on the atmosphere, including the risks of disturbing the ionospheric equilibrium;
- effects of the microwave beams on the biosphere, including interactions with living organisms in receiving areas;
- radiofrequency interference;
- noise pollution resulting from the launching of the satellites;
- pollution of the atmosphere from the satellite launching operations;
- land use, etc.

GLOBAL REPERCUSSIONS OF INTENSIVE SOLAR ENERGY USE

Land use and Aesthetic Impacts

One of the most frequent objections to solar energy is that it uses up too much land. It is true that a 1 megawatt central receiver type solar power plant occupies about 3 to 4 hectares, whereas a 1,000 megawatt nuclear plant takes up only 50 hectares. However, it has been calculated that a tenth of the total energy needs of France in the year 2020 could be met by various types of solar facilities occupying a total land area of 480 km^2, assuming a mean energy conversion efficiency of 20%. This is only 8 m^2 per inhabitant, and should be compared with the 4 m^2 per inhabitant that would be necessary to double the length of the existing French motorway system (Rodot, 1979).

Pressure groups and lobbies have been created in some countries to promote projects in which conventional and nuclear energy source would be entirely replaced by renewable energy sources e.g., the Belleville group's ALTER project (de Closets, 1978). Studies have shown that direct and indirect solar energy use could probably satisfy the world's total energy needs if this were absolutely essential (Mitre Corporation, 1975). Several techniques of energy conversion would require further development, including solar thermal collectors, photovoltaic and photogalvanic devices, wind-powered generators, ocean thermal power plants, and hydropower plants. The total area required for collectors and other facilities would be very large, estimates ranging up to 8 million square kilometres of land and 11 million square kilometres of ocean. Such a proliferation of solar equipment would seriously modify and degrade landscapes, and require abandoning of traditional architectural styles. Ensuring that changes of the magnitude in question take place harmoniously and are acceptable to the populations concerned would be by no means easy, and this is perhaps an ideal (if rather vast) field of study for sociologists, economists, industrial designers, and the like.

Climatic Change

One of the questions being asked by those giving serious thought to the long-term future of humanity is whether mankind is capable of so affecting the global climate as to bring about conditions causing widespread human disasters or even its own demise. In this respect, it must be noted that human influences on the global heat balance have to date arisen from 5 basic sources:

- Changes to the reflectivity or absorptivity of the earth (i.e., to its albedo). Many human activities such as cutting down forests, plowing up grasslands and building cities affect the albedo and result in net heating of the planet.
- Irrigation. Although it can cause local cooling, the ultimate net effect of irrigation is heating of the planet, since evaporation of irrigation water absorbs solar radiation.
- Particulate emissions. Man-made fires and industry release soot, ashes, etc. into the atmosphere, and this probably produces net cooling due to reflection and scattering of sunlight back into space by the emitted particles (wind-blown dust and volcanic eruptions also play a major role here).
- Carbon dioxide emissions from combustion of fossil fuels. Atmospheric CO_2 absorbs radiation and thus produces net heating of the earth.
- Direct release of heat. All uses of stored energy, whether fossil, nuclear or (to a lesser extent) geothermal, result in net heating of the planet.

Only the first and the last of the above are applicable to intensive use of solar energy.

As explained earlier, global change to the albedo would be only slight, even if significant local changes occur close to collector and heliostat fields. However, if it were decided to satisfy all or almost all of the world's energy needs with renewable energy sources, then the net environmental impact of the gigantic land-based collector arrays and ocean power plants necessary would significantly redistribute the solar energy received on earth, and this could have major climatic consequences.

With respect to heat releases, intensive capture of the solar radiation received on earth would not change the global energy balance, but only modify the local energy distribution. In comparison, many detailed studies have shown that a large increase in the amount of carbon dioxide in the atmosphere (e.g., due to burning fossil fuels) might considerably change the thermal equilibrium of the planet, leading to a perhaps catastrophic global temperature rise (Peyches, 1977). Since the end of the 19th century, the amount of carbon dioxide in the atmosphere has increased by 10 to 15%, and this has raised the planet's temperature by about 0·2°C. The temperature will increase by another 0·3°C by the year 2,000, or 2°C if the present rate of atmospheric CO_2 build-up (0·2% per year) were to double. Substituting solar energy for fossil fuels is one way of catering for rising energy demand while limiting the amount of CO_2 released into the atmosphere, and thus overcoming the dangers of an excessive temperature rise (e.g., melting of the polar ice caps, causing a rise in the level of the oceans, and submerging of large areas of land).

A study commissioned in 1973 by the National Science Foundations in the United States (Mitre Corporation, 1975) calculated the increase in the planet's temperature, both globally and at the poles, for the three basic scenarios of a high, medium or low growth in energy consumption, with two options per scenario:

- option A: increased consumption satisfied with fossil fuels,
- option B: intensive use of solar energy.

The results of this study are summarized in Table 3.1 (Breniere, 1980):

Table 3.1 Increase in the Planet's Temperature for Scenarios of High, Medium and Low Growth in Energy Consumption

	Average temperature increase (°C)	
	GLOBAL	AT POLES
Scenario I Option A	2–3	10 or more
Scenario I Option B	1*	2–3*
Scenario II Option A	1·2–1·3*	2–5*
Scenario II Option B	1*	2–3*
Scenario III Option A	0·5*	1*
Scenario III Option B	0·5*	1*

The predominant cause of these temperature increases is the CO_2 concentration in the atmosphere due to combustion of fossil fuels.

It is clear from these data that the use of fossil fuels must be limited in the future. The atmospheric CO_2 concentration should not be allowed to exceed 400 to 420 ppm, and consumption of "stored" energy should be limited to a total of 3×10^{18} kJ per year or less, to avoid significant temperature rises at the poles. Intensive use of solar energy and curtailment of fossil fuel utilization would make possible a total energy consumption of 20×10^{18} kJ without significant global or polar temperature increases.

Another advantage of solar power plants over nuclear and fossil fuel plants is the question of water vapour and heat discharge to the atmosphere. Future very large non-solar power plants might release several cubic metres per second of water vapour from their cooling towers, and this could lead to the formation of cumulo-nimbus clouds several kilometres in size within a very short time; the local heat release could be equivalent to the solar radiation received in summer over a surface of 3,000 km^2.

THE FUTURE

The prediction and evaluation of future possibilities necessarily involve a great deal of guesswork, especially in our world of ultra-rapid change. All that can be said with certainty is that solar energy has a role to play in satisfying the energy needs of mankind, but will not for a long time (if ever) be able to take the place of more conventional sources of energy, except for some limited applications (Cintra do Prado, 1979). One of the fundamental handicaps of solar energy is the rather low level of energy efficiency practically obtainable (from a few per cent to 15–20 per cent, according to the various technologies available).

Fossil fuels will continue to satisfy a large proportion of total energy requirements for many more years, even though gradually reserved for uses where they are for the moment irreplaceable (e.g., motor cars). Nuclear energy use will continue to grow, with maturing of advanced forms of nuclear reactors such as high-temperature reactors, fast breeder reactors and fusion reactors.

For these reasons, solar energy utilization will not have major regional or global consequences for the environment within the foreseeable future, and any local impacts could be easily mitigated by suitable legal and regulatory measures. The relatively minor environmental inconveniences of solar energy are, however, more than outweighed by its beneficial effects. For arid regions, solar energy can help solve social and environmental problems by reducing deforestation, stopping or slowing-down migration of rural populations to urban centres, improving communications, making electricity and other forms of energy available locally for isolated communities, preserving food and medical products, and so on. For the industrialized nations, solar energy use can limit or reduce fossil fuel burning for heating and electricity generation, decrease national dependence on imported energy, etc., and the emergence of new solar industries will help create jobs.

One of the major future applications of solar energy may be production of hydrogen by photolysis of water or electrolytic methods. Indeed, some authors have imagined that the civilizations of the future will employ hydrogen as their principal source of energy (Gibrat, 1974). The consequences of such a "hydrogen economy" for our ways of life, national economies and patterns of trade, and international relationships are easy to imagine. Of course, in reality (as opposed to our visions), there are many practical obstacles to this and many other revolutionary changes. Thus, producing massive amounts of hydrogen using equally massive solar energy plants might cause unacceptable modifications to the world's climate, due to multiple local changes in the albedo, and the global redistribution of the received solar energy, even though the climatic impact of solar energy would almost certainly be less than for other potential sources of heat. Again, this illustrates that two of the key deciding factors for the future "choice" between different sources of energy will probably be their impacts on global climate, and their local micro-climatic impacts directly related to the installed power capacity at each particular site.

References

AFEDES (1980): Guide des Installations Solaires dans l'Habitat. Mars.

Alexandrof, G. and A. Liebard (1979): L'Habitat solaire, comment? Editions Apogee, Paris.

Balligand, P. *et al.* (1979): Serres en agriculture, WMO, MSE/Doc. 4, Madrid, 23 Janvier.

Balligand, P. *et al.* (1979): Serres à dessalement solaire. Proceedings of the International Congress on Desalination and Water re-use. Nice.

Bloss, W. *et al.* (1979): Session for power production. Solar Energy for Development, Varese.

Breniere, F. (1980): Etude des impacts de l'energie solaire sur l'environnement. Institut Francais de l'Energie, Paris.

Caputo, R. S. (1977): Solar power plants, dark horse in the energy stable. Bull. Atomic Scientists. May.

Channabasappa, K. C. (1979): Proceedings of the International Congress on Desalination and Water re-use. Nice.

Chouleur, G. (1975): Les maisons solaires d'ARAMON. Documentation Electricite de France.

Cintra do Prado, L. (1979): Energia Solar. Uma realidade atual. Mundo econico, politico e social, Sao Paulo, Dezembro.

Clemot, M. (1978): Solar pumping systems. Annals of Arid Zones, New Delhi, 1975; Soltech, Barhein.

de Closets, F. (1978): La France Solaire du XXIe Siecle. Sciences et Avenir, No. 374, Avril.

Collection Environnement (1974): Rapport sur les relations entre le secteur de l'Energie et l'Environnement dans une perspective a long terme. Collec. Envi r. No. 35, La Documentation Francaise, Paris.

Colomes, J. and S. Roland (1977): Resultats et perspectives d'exploitation des stations heliometriques SHA.K7. XVIe Recontre Internationale de la Comples, Tunis.

Congrès solaire international (1979): Utilisation de la chaleur solaire dans l'industrie et l'agriculture. Nice.

D.O.E. (1977): Solar thermal Power Systems. DOE. EDP. 0004. US Department of Energy, Washington D.C.

Durand, H. (1979): Power Production, Key note speech. Solar Energy for development, Varese.

Durand, H. (1979): Utilisation de la chaleur solaire dans l'industrie et l'agriculture. Congres International, Nice.

EEC (1979): Solar Energy for Development. EEC Conference, Varese, Avril.

EPA (1976): Potential Environmental Impacts of Heating and Cooling systems EPA. 600/7.76.014. U.S. Environmental Protection Agency, Washington D.C.

Etievant, C. and F. Pharabod (1977): Le projet THEM. L'industrie du petrole, N. 481, Avril.

Forster, G. H. and R. M. Peart (1976): Agriculture Information Bulletin No. 401, Nov.

Gibrat, R. (1974): L'Energie en l'Energie en l'an 2,000. Communication a l'Academie des Sciences Morales et Politiques. Paris.

Gicquel, R. (1978): Le comportement des capteurs solairs dans un systeme. Le Moniteur, cahier technique, December.

Girardier, J. P. (1973): Communication, Congres International "Le Soleil au Service de l'Homme". UNESCO, Paris.

Glaser, P. E. (1979): La mise au point des satellites a energie solaire. Revue de l'Energie, No. 313, Mars.

Hoda, M. M. (1978): Solar Cookers. ISEC International Solar Energy Congress, New Delhi, January.

Liebard, A. (1980): Les capteurs a ruissellements; Guide des installations solaires dans l'habitat. AFEDES, pp. 54-57.

Manjarrez, P. and M. Galvan (1979): Solar multistage flash evaporation as a solar energy application on desalination processes. Proceedings of the International Congress on Desalination and Water re-use. Nice.

Maurel, A. (1979): Dessalement et Energies Nouvelles. Proceedings of the International Congress on Desalination and Water re-use. Nice.

Mitre Corporation (1975): Energy Use and Climate, Possible Effects of Using Solar Energy instead of Stored Energy. M 74-66 Rev. 1 NSF. RA. N. 74. 185 – April.

Mitre Corporation (1979): Solar Industrial Processes. The Mitre Corporation, MP 79 W 00005, March.

Morse, M. N. (1977): Solar heating as a major source of energy for Australia. 10th World Energy Conference, Istanbul, paper 4. 2-3.

Muthuveerapan, V. V. (1979): Solar heat for Agriculture. Proceedings of the International Conference EEC, Varese.

NAS (1976): Energy for rural development. National Academy of Science, Washington, D.C.

OTA (1978): Application of Solar Technology to today's energy needs. Vol. 1 Office of Technology Assessment. Congress of U.S.A., Washington D.C.

Palz, W. (1977): Solar Electricity, an economic approach to solar energy. UNESCO, Butterworth Publishers, London.

Pattanayak, S. *et al.* (1978): Continuous solar grain dryer. ISES International Solar Energy Congress, New Delhi.

Peri, G. (1977): Les capteurs solaires plans. Cahier de l'AFEDES No. 4.

Perrin de Brichambault ch. and G. Lambolley (1975): Le rayonnement solaire au sol et ses mesures, Cahier de l'AFEDES No. 1 Juillet.

Peyches, Y. (1977): Repercussions climatiques de l'utilisation des energies primaires. Cas de l'energie solaire. Rev. Gen. de Thermique, No. 184, Avril.

Pheline, J. (1979): Energie solaire et production decentralisee d'electricite. Annales des Mines, Paris, Avril.

Photovoltaic Specialists Conference (1980): Electricity from photovoltaic cells; Status of low cost solar Array project. 14th photovoltaic specialist conference, January.

Presente, G. (1977): Methodologie applicable a l'urabisme et a l'habitat pour l'utilisation de l'energie solaire. Journees solaires de Nice, Avril.

Rodot, M. and P. Vasseur (1977): Cahier No. 2 de l'AFEDES, Paris.

Rodot, H. (1979): Energie Solaire et Environnement. Rapport interne, Ministere de l'Environnement et du Cadre de Vie.

Torrenti, R. *et al.* (1976): Dimensionnement d'un bassin de stockage intersaisonnier. Comples, 2e semestre.

Trombe, F. and J. Michel (1977): Utilisation de l'Energie Solaire dans l'habitat. Revue chauffage, ventilation, conditionnement, Mai.

Trombe, F. and A. Le Phat Vinh (1973): Le four solaire de 1,000 KW du Centre National de la Recherche Scientifique. Communication au Congres International "Le Soleil au Service de l'Homme". UNESCO, Paris, Juillet.

UNEP (1979): Solar 2,000. UNEP Report 8. United Nations Environment Programme, Nairobi.

UNIDO (1978): Technology for solar energy utilization. Development and Transfer of technology series No. 5; UNIDO, Vienna.

U.S. Department of Energy (1978): Status report, Washington, D.C.

Walton, J. D. *et al.* (1978): Survey of solar powered irrigation pumps, Solar cookers, and wood burning stoves for use in sub-sahara Africa. Georgia Institute Tech. Atlanta, Georgia.

Young, L. L. (1979): Health and Safety hazards associated with solar concentration systems. Solar Energy, Great Britain.

CHAPTER 4

Wind Energy

BENT SØRENSEN
Professor of Physics, Roskilde University Centre,
Roskilde, Denmark

Fig. 4 Windmill on Grazing Land (photo: FAO)

INTRODUCTION

WIND energy may be utilized by conversion into linear or rotational motion. An example of conversion into linear motion is the use of sails to propel ships. Conversion into rotational motion may be performed by stationary wind turbines. The rotating shaft power may be further transformed into pumping motion (for water pumping), electric power (by an electric generator) or heat (e.g. by hydraulic dissipation of the mechanical energy). Direct production of electric power is possible by using a wind driven ion generator. There is a variety of conversion devices, ranging from the classical sail types, windmills and panemones to modern aerodynamically shaped horizontal or vertical axis converters and advanced concepts. A technical description of the conversion methods and application of devices and systems based on wind energy is outside the scope of this book (but may be found elsewhere, e.g. in Sørensen, 1979). Here focus is made on the environmental impacts of existing or proposed wind energy conversion devices, with emphasis on free stream turbines and occasional reference to sailships and advanced stationary converters.

In assessing the environmental impacts of wind energy, two further aspects have to be considered: the type of use and the support system required beyond the conversion device itself. The type of use is important for determining the range of alternative energy generation methods, to which the impacts of wind energy conversion should be compared. For rural irrigation purposes the alternative may be diesel-driven pumps, for main grid electricity production the alternatives may be coal or nuclear fuel use in central power plants, and so on. The specifications of the use also determines the support system required. If power has to be supplied on demand at any time, energy storage facilities or back-up systems that can be regulated at will have to be part of the supply system. The total assessment of environmental impacts should thus include manufacture and operation/maintenance of equipment, for conversion, storage and transmission as required by the actual application. In some cases, the impact of some components will be independent of the energy source used, and in such cases an assessment of alternative conversion methods may be performed in isolation, by assuming that the remaining parts of the system (e.g. transmission facilities etc.) will be identical for all the alternatives.

SAFETY HAZARDS

A number of situations may be imagined, in which wind energy technology could affect the physical safety of human beings. Falls from high locations could be associated with the tower constructions characterizing many wind conversion systems. The accident probability is likely to be greater during tower construction for large stationary turbines, during maintenance that has to be carried out at high altitudes, and for sailships during manual setting from high masts. Most currently constructed wind conversion devices are made in such a way, that dangerous climbing is eliminated during routine operation (automatic setting of sails on sailships, maintenance-free constructions of stationary converters, inside stair access to nacelle of large turbines, etc.) However, during construction and repair, hazards may still exist. They are of a nature common to most structures (tall buildings, masts for overhead transmission lines, and so on). Therefore the nature of the hazards is well-known, and measures to reduce the risk below given standards are generally known. However, in traditional construction industries, norms have been developed over long periods of time, and it should be made certain, that wind conversion devices are subject to appropriate standards and codes, or that standards will be set to ensure that such an emerging industry will not exploit loopholes in existing regulation primarily for economic profit.

Consider now failure of rotor-type converters in a little more detail. Failure may occur in each component of the wind energy system: rotor blades, shaft, transmission or tower (Fig. 4.1). Causes include fatigue of materials, corrosion, fretting or moisture ingestion, depending on the nature of material used (e.g. metals, wood, etc.). A number of loads on the converter may contribute to fatigue failure: different wind velocities experienced by different parts of the turbine; changing speed and direction of wind; wind turbulence and gusts; accelerations and stresses on components, including aeroelastic and resonance effects; tower-rotor or blade-blade interactions; gyroscopic and centrifugal loads at speed changes and start or stop. Excessive loads may particularly arise in cases of failing controls, e.g. leading to overspeed.

The engineering design principle for wind energy converters is safety throughout lifetime. Ensuring this requires not only a static calculation of the construction, but also a dynamical, aeroelastic calculation for the turbine under normal operation as well as under various abnormal conditions (brake failure, etc.). For a horizontal axis turbine, the main fatigue-producing motions of the blades are the flapping modes (in and out of the plane of rotation) and the lead-lag modes (in the plane of rotation). Fig. 4.2 shows an example of the calculated variations in flapping and lead-lag amplitudes during one revolution for a blade on a two-bladed horizontal axis rotor of diameter 38m (Kottapalli *et al.*, 1978). The model can include the effect of blade-tower interactions, which is seen to affect the maximum amplitude of the flapping mode considerably. For a discussion and comparison of a number of aeroelastic models see e.g. Spera (1978), Friedman (1978) and Raab (1980). The latter one includes the effect of the turbulent component of the wind, which in some cases can be substantial. Generally speaking, it is possible numerically to estimate the behaviour of horizontal axis rotors under normal operation with sufficient accuracy to allow safe distance to material failure situations, as it is similarly possible for the tower construction. The same applies to a number of cross-wind turbines, although the computational experience is smaller than for horizontal-axis machines.

Fig. 4.1 Tower failure of commercial wind energy converter in the 20 kW range.

Fig. 4.2 Calculated amplitudes of flap and lag during one revolution, with and without tower shadow (Kottapalli *et al.*, 1978)

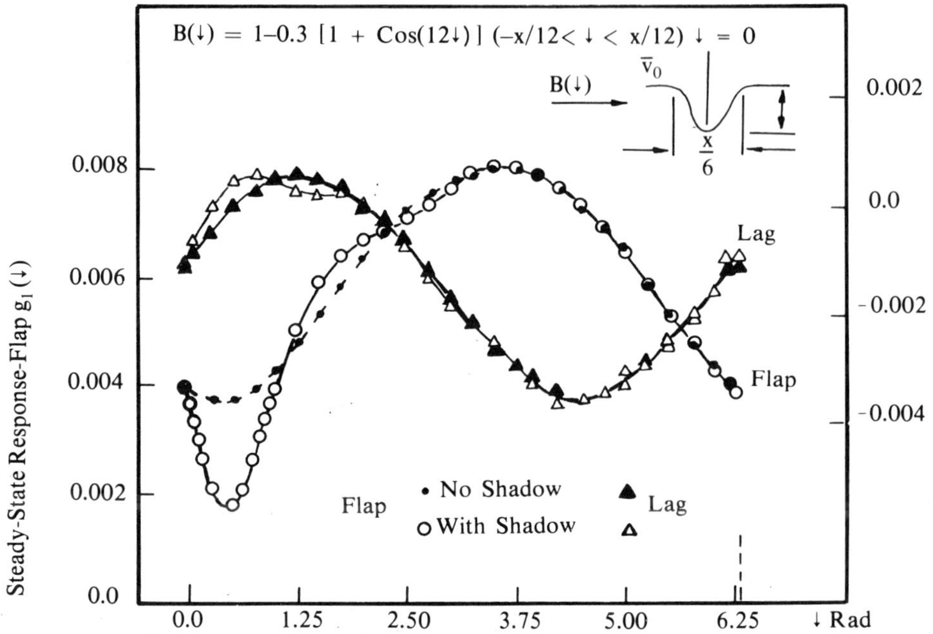

More difficult is the anticipation of failure modes caused by abnormal situations, since the details of such events can rarely be known in detail. As mentioned a failure in the brake or control system may fall into this category, and also events caused by external factors, such as strikes by lightning, impact of various objects like birds, airplanes, etc. In many cases the only defence against such events of failure, e.g. causing pieces of rotor blades to be expelled and travelling distances of hundreds of metres, is to site the wind energy converters away from dwellings and areas frequented by many people. This precaution is often easy to fulfill for large, electricity-producing converters, whereas the users of small machines often prefer to have the wind energy converter near to their house or farm. The approach here should be to minimize the risk of external failure causes, and to insist, e.g. by regulations, that the rotor is "failsafe" under all conditions not involving external causes of unknown character. Extreme winds are of course part of the design basis, as are extreme lightning intensities for which ground-connected rods of sufficient current-carrying potential must be incorporated in the tower and wings.

With respect to the materials problems under the general heading "fatigue", the balance of failsafe construction and economic penalty may be improved by proper inspection techniques, that acknowledge the occurrence of cracks but detect them and allow repair before they lead to failure (Bridson and Worthington, 1980).

Presently, the prime candidates for storage* in connection with electricity-producing wind converters are batteries and hydro systems with reservoirs (Sorensen, 1980a). These systems may also involve safety hazards, such as dam failures for hydro systems. Approaches to minimizing risk include siting or re-siting of dams and significant human settlements in such a way that no populated area will be flooded in case of dam failure, or alternatively an inspection, warning and evacuation system that will work. Future energy storage forms will have their particular sets of safety-related environmental problems. Hydrogen storage, for example, may involve a gas explosion risk, that may require a separation of the system, so that the bulk of the hydrogen is stored far from populated areas, and only small amounts are transmitted to the load centres for re-conversion to electric power, e.g. in fuel cells, at any given time.

HEALTH HAZARDS

Few health effects are associated with the operation of wind turbines, which does not require use of toxic substances beyond lubrication oils and similar small scale usage of moderately harmful chemicals. The main risks of chemically induced substances of possible toxicity is likely to be associated with the construction phase: use of epoxy and glues in blade manufacture e.g. glass fibre or laminated wooden materials, dust and chemicals introduced into the air during erection of concrete towers, and the health hazards associated with the manufacture of steel used in transmission and generator parts, as well as for tower and blades in some constructions.

Noise

The rotor type wind energy converters may give rise to noise associated with the interaction between the moving and static parts and the wind, as well as mechanical noise in the power transmission and conversion machinery gear-boxes, generators, etc. The mechanical noise associated with power transmission may present a health hazard for persons working in the nacelle of large turbines during operation, and to a lesser extent to people on the ground. In the case of noise exposure to the general public, smaller wind energy generators may be of more concern, since they are typically sited close to dwellings, while large converters producing electric power for a general grid system are typically sited at considerable distances from areas occupied by people. In any case, sound insulation of transmission machinery is feasible, and the present standards for

* see Chapter 7.

occupational and general exposure, which differ from one country to another, can be met in all cases. Typical limiting standards are 85–90 dB(A)*, while recommended guideline noise levels are around 75 dB(A) for an 8 hour working day (Keast, 1978).

For external noise, i.e. noise experienced by a person on the ground outside the premises of the energy conversion plant, standards recommended are from 70 dB(A) in industrial areas and down to 35 dB(A) for residential and recreational areas during night-time (Danish Dept. of the Environment, 1974). The noise from wind turbines must be confined to the night-time values if operated during night. If operation during certain night hours is prohibited, the total energy production decreases, and measures to keep the noise down so that the machine can be operated would usually be economically warranted.

Noise levels not presenting an established health hazard may nevertheless be annoying to people. Fig. 4.3 gives an impression of intervals of noise levels perceived as "highly annoying" by various percentages of people surveyed in a number of situations. Noise levels that will cause most sleeping people to wake up range from 40 to 90 dB(A), depending on stage of sleep and individual differences. Some ten per cent may even wake up from light sleep at brief sounds in the range from 0 to 30 dB(A) (Keast, 1978). Conversation at normal level of voice is possible at a noise level of 40 dB(A) over distances up to 20m. This distance decreases linearly with noise level in dB(A), and reaches zero at 80 dB(A). Prolonged exposure to noise causes a transient, or at worst permanent, increase in the threshold for perceiving sound. For 24 hours exposure at 82 dB(A), the threshold shift as measured two minutes after exposure is 35 dB, and for noise levels above 82 dB(A) for prolonged periods the possibility of acoustic trauma exists (Keast, 1978).

In contrast to the noise caused by mechanical parts of power transport and conditioning, which can be brought below any conceivable standards by proper design, the noise created by the interaction between the turbine and the wind itself is of a more fundamental character and it cannot be arbitrarily reduced. This "aerodynamic noise" has several components. First there is the static noise caused by the interaction between the wind and the tower and rotor parts even in a condition of no movement. This noise is similar to the one found in connection with all structures or obstacles like houses, trees, masts, etc., but it may depend strongly on the type of wind converter construction, such as use of large-diameter hollow tower, slender pole-type tower of metal or concrete, with rough or smooth surface, or metal grid tower constructions.

Secondly, noise caused by the motion of the rotor must be considered. It consists in part of noise of frequency equal to the rotational frequency or multiples of this, and in part of broadband noise caused by the interaction between the turbulent component in the wind and the rotor blades. There are several mechanisms linked to the rotational frequency, including the periodic

* dB(A) is a measure of noise on a logarithmic (decibel) scale simulating the human perception of noise levels in the audible range by use of frequency-dependent weighting factor, the "A-filter".

shedding of vortices from the air foils and rotor-tower interactions, particularly if the rotor is placed downwind.

Fig. 4.3 Summary of 12 social surveys of noise annoyance (based on Keast, 1978)

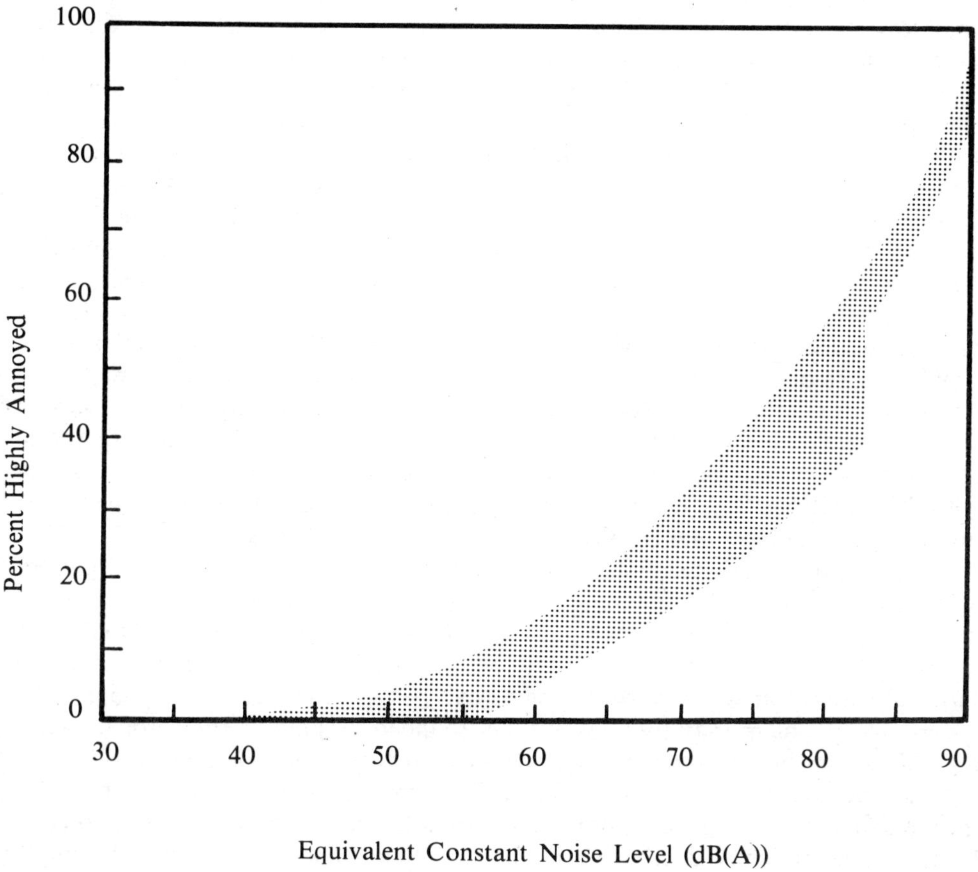

Several measurements of noise levels near existing wind turbines have been performed. At a distance of 50m from a Danish turbine from 1957 (rotor diameter 24m, fixed rotational frequency 30·2 r.p.m.), the aerodynamic noise at a windspeed of 15 m/s was 10 dB higher than the noise level associated with the transmission system (Danish Dept. of Energy, 1978). The latter was 50-52 dB(A) due to an inappropriate gear construction. For the more recent US Mod-O construction of 1975, the total noise was found to be 20 dB(A) at the foot of the tower, reducing to 10 dB(A) at a distance of 90m (Rogers *et al.*, 1977; Sørensen,

1980b). This machine has a rotor diameter of 37m and a rotational frequency of 40 r.p.m. The noise measurements were performed at wind speeds between 5 and 8 m/s. It is concluded that audible noise is not a problem in connection with large wind generators. For smaller machines near farms or other houses, there would typically be other structures around, which produce aerodynamic noise, e.g. trees, and the main concern would be to keep noise from transmission to an acceptable level.

In many cases, wind turbines create the largest noise intensity outside the audible frequency range. Infrasound, i.e. sound of frequencies lower than the lower limit of human perception, is emitted in association with the vortices shed by the blade tips and to some extent roots of a wind turbine rotor. The dominant infrasound contribution, however, is from the interaction of vortices formed by the passage of wind around the tower structure with passing blades, in the case of downwind rotor mounting. For the US Mod-O machine, signals in the range 90-100 dB and of frequencies near 1 Hz have been measured close to the turbine (Rogers *et al.,* 1977). For the larger Mod-1 machine, which also has the rotor downwind and has a steel lattice tower construction, infrasound intensities at a house located over 500m away has caused annoyance to such an extent, that night-time operation has been restricted. On the other hand, no infrasound problems has occurred for the similar size machine at Tvind in Denmark, which has dwellings situated 50-100m away from the turbine, and which is also a downwind machine. This indicates, that tower structure is a major factor in reducing the infrasound intensity. The machine at Tvind has a concrete tower with three sections of different diameter, such that the unfolding vortex sheets will interfere and lose coherence after short travelling distances.

The health impacts of infrasound are somewhat different from those of audible noise. Generally speaking, the thresholds for annoyance as well as for health hazards are about 40 dB higher for the frequency range 0·1–5 Hz (Rogers *et al.,* 1977). Thus, it should be possible to construct downwind horizontal axis turbines without annoyance problems as can be noted in case of the machine at Tvind. The problem may by substantially reduced by placing the rotor upwind.

Noise problems are usually addressed in terms of distance between the turbine and areas occupied by people. Therefore, it is important to know the mechanisms of sound propagation. While sound propagation in homogeneous air leads to quadratically decreasing intensity with distance, the effects of ground reflection and transmission through obstacles like walls, vegetation, etc may substantially modify the propagation. The same is true for inhomogeneity of the atmosphere, such as the one associated with temperature and wind speed profiles and their changes with times. Inhoniogeneity usually causes a reflection of the sound. During night-time inversions, i.e. positive temperature-height gradients, this ceiling may disappear. The complexity of sound propagation should be taken into consideration when locating wind machines.

In comparison with alternative ways of energy production, the use of wind energy may dramatically reduce the noise production, as for example when replacing a diesel engine by sails in a ship.

INTERFERENCE WITH TELECOMMUNICATION

Wind energy converters may cause distortion of telecommunication signals, either due to reflections on the structure or due to modulation caused by the electrical properties of moving parts.

Television interference may be discussed in terms of two parameters. One is the scattering cross section or effective area of the turbine blades with respect to scattering of electromagnetic waves. The other one is a threshold modulation or maximum amplitude of the interfering signal, which will be judged as acceptable (Sengupta and Senior, 1978). The interference between the scattered part of the signal and the direct one is responsible for the observed distortion of audio or video reception.

The scattering, which is dominated by specular reflection from the broad face of the blades, corresponds to effective areas of 8–25% of the rotor swept area, depending on blade material. The highest value is for metal blades, the lowest one for composite materials such as glass fiber. The effect of adding metal strips to blades made of composite materials, for lightning protection, is of the order of 20% increase in the effective scattering area. These estimates are deduced from measurements performed on the US Mod-O by Sengupta and Senior (1978).

The effect of interference with a rotating object of a given radio or TV signal is an amplitude modulation showing up as a repetitive pulsed distortion. All other things being equal, the amplitude of the modulation increases with frequency, thus being small for FM radio signals and large for the highest (UHF) TV channels. The distortion depends on the directions of turbine blades, e.g. yaw angle, and the relative directions to turbine and sender as seen from the point of reception. The signal path may furthermore be indirect in mountain regions or due to other obstacles. Measurements performed 150m from the US Mod-O has led to the establishment of a maximum amplitude of the interfering signal as being 2·6 dB (15%) in the most critical direction and for TV channel 50. In other directions or for lower frequencies the modulation threshold is higher. For instance, a value of 16 dB was found for a healthy FM signal (signal to noise ratio over 15 dB).

The aim of the study by Sengupta and Senior was to set standards for calculating the extent of interference zones for a given wind turbine construction. If houses are located within this zone, a feasible solution may be to provide them with cable connections to aerials placed outside the interference zone. This solution has been used by the Danish Department of Energy (1978), which found an interference zone extending to ten rotor diameters from the turbine, for the machine mentioned in the previous section.

The presence of wind energy converters may interfere with other forms of telecommunication. Examples are the microwave communication links used by telephone companies and the navigational systems used by commercial aircraft authorities. Sengupta and Senior (1978) studied these problems and found that, in the case of current navigational systems the interference is maximum when the rotor is not moving, and decreases when the blades are turning.

IMPACTS ON ECOSYSTEMS

The impacts of wind energy conversion systems on ecosystems may result from the extraction of materials used in the construction of equipment, from the operation of the wind converter itself and from other parts of the system, such as transmission and energy storage facilities. Free-stream wind turbines are not particularly intensive in material use, as compared with other energy conversion equipment. For this reason, no specific discussion of the impact of material extraction will be attempted here.

The operating wind turbine may present a hazard for birds and airborne insects. Some windy sites, e.g. near seashores or large lakes, are of vital importance for migrating bird species. Migrating songbirds normally fly at a: altitude of 150–450m, while waterfowl and shorebirds fly at higher altitude. (Rogers, *et al.,* 1977). Many nocturnal migrant birds have been killed by tall buildings, towers, transmission lines and other structures without moving parts. A major factor in the killing of birds has been the lights sometimes installed on top of towers, structures, etc., since light may strongly distract certain birds from avoiding collision courses. Experimentation is underway to determine the colour and intermittancy pattern of the light that will minimise such disorientation.

It is likely that a stationary wind turbine presents a higher risk than a rotating one, because the birds may sense the motion and take evasive action. A survey around the first US Mod-O wind turbine, maximum height of about 50m, showed that one bird was probably killed by the rotor tower and two probably killed by the nearby meteorological tower; the turbine was not operating during most of the four migration seasons covered by the survey. Bird-kills by a 150m cooling tower of a nuclear power plant in the same State amounted to between 44 and 103 per season (Rogers *et al.,* 1977). It, therefore, appears that the height and solidity are important factors in determing the risk to birds.

The Mod-O study further investigated the behaviour of the birds approaching the turbine. The migration traffic rate ranged from zero during off-season having precipitation to 10·6 birds per hour and per metre crossed perpendicular to the direction of flight during clear night in spring migration season. The average for a clear night was 3·4/h/m. Of a series of birds heading towards the rotating turbine, six out of nine took evasive action and avoided a course penetrating the area swept by the rotor. Three other birds headed on a straight course between the turbine blades without any incident. This supports the theory that many birds will sense the danger sufficiently in advance, and further stresses the fact that the blades of large wind turbines are slowly moving objects, which may not present a hazard even for birds flying between them. The situation is somewhat different for small wind turbines, which if sited in suburban regions may present a danger to garden birds. This study also indicates, that the open nature of the obstacle may give better chances for evasive action than the solid obstacle of the cooling tower mentioned above.

Records of bird killing, e.g. in connection with light towers, has shown very irregular patterns, with major occurrences at intervals in specific situations of high migration rate, heavy fog or mist and certain wind directions. These conditions prevailed during the Mod-O study but no strikes were observed. The migration rate was 4·9/h/m and the turbine was operating during 4 hours. Flying insect populations were released near such turbine, but no effect could be deduced (Rogers, *et al.,* 1977).

The highest potential ecosystem impact of wind energy systems will be associated with the energy storage systems. Likely storage systems could be batteries, pumped hydro installations, back-up systems regulated in anti-phase with the wind turbines, e.g. hydro turbines or fossil fuel generating systems. Batteries have a limited lifetime, and discarded batteries constitute a form of chemical waste, which must be properly managed. The environmental problems associated with fossil fuel-based systems are well-known (see, for example, UNEP, 1981). Thus, the use of fossil fuel generators as back-up systems should be seen in the context of a systematically declining dependence on fossil fuels as the main source of energy.

Hydropower* and its use in combination with wind energy conversion is a different problem, because hydro energy is a renewable source which may find permanent place in future energy systems, due to its versatility as storage or back-up for intermittent sources in particular the short time needed for switching and the reservoir options for truly long-term energy storage. It is, therefore, very important to consider the environmental impacts of hydro installations.

WORK ENVIRONMENT

The working conditions for operators of wind energy systems are generally good, with respect to variations of tasks and possibility of participation in decision making and function in groups of modest size. Repair and maintenance services offer a generally varied and challenging work situation. Typically, one team would be servicing a number of units installed in arrays or more dispersed. A good fraction of the work is likely to be outdoor activities, and some repair work on outside parts of tower or blades may involve unpleasant climbing and high altitude stays possibly under adverse weather conditions. Safety precautions would be similar to those used by scaffold construction workers.

*See Chapter 6

VISUAL IMPACTS

The visual impacts of wind energy systems result from the converters, from transmission systems and from support facilities such as energy storage systems. Transmission lines in connection with electricity producing converters may be required in addition to those associated with the transmission of power to load locations. This is true, if arrays of wind energy converters are placed in high wind areas far from power consuming areas. Conventional steam power plants also have this problem, because they require large quantities of cooling water and, therefore, are typically sited at seashores or major rivers, mostly away from load centres. There is considerable motion to replace overhead transmission lines by underground cables, in populated or recreational areas. Energy storage systems may also present a visual impact, such as the case of a large dam and the landscape changes induced by hydro reservoir construction.

The wind energy converter itself may possess artistic or functional beauty, as witnessed by many sailships and windmills. The designers of modern wind energy converters have in some cases considered this aspect. If large numbers of wind turbines become accepted as part of our visual environment, their design and incorporation in the physical surroundings are important factors. Several studies have been made or are under way, which aim at the detailed assessment of the visual impacts of wind turbine arrays. Figs. 4.4 and 4.5 give an example of such a study (Birk Nielsen, 1980). 161 wind energy converters have been placed in arrays near the Baltic coast of Denmark (Fig. 4.4), on reclaimed land currently used for agriculture. This land use would not be altered by the wind turbines, assumed to have a rotor diameter of 50m and a minimum spacing of 7 rotor diameters. Fig. 4.5 illustrates the view of the wind turbines; the conclusion is that the converters have a substantial impact on the landscape, but such impact need not be considered negative. It can indeed help to underline the characteristics of the landscape.

Siting of wind turbines offshore has also been considered not only to minimize the visual impacts but also to take advantage of better wind conditions at sea. Offshore locations will enable the utilization of wind energy where land areas are so densely populated that no suitable sites can be found.

MICROCLIMATIC IMPACTS

Microclimatic modifications may occur in areas where wind energy is extensively exploited. The energy extraction from wind is accompanied by a slowing down of the wind. This may lead to changes in precipitation and evapotranspiration patterns downwind, and conceivably to changes in temperature and concentration of minor atmospheric constituents as well.

Fig. 4.4 Siting proposal for 161 wind turbines (indicated by dots) near Rodby, Denmark

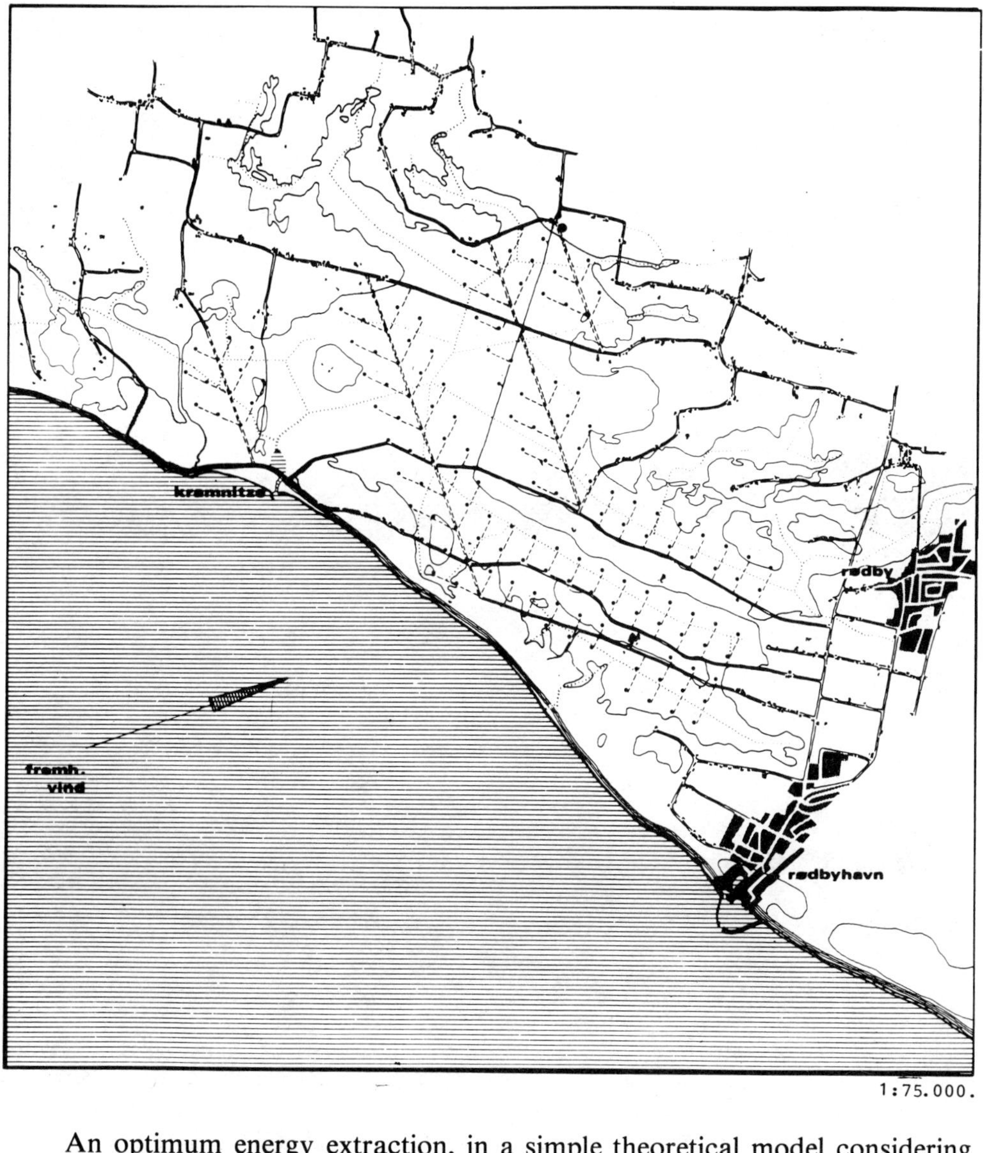

An optimum energy extraction, in a simple theoretical model considering streamwise flow only, is achieved through a reduction of the incident wind speed, downwind from the rotor, to one third (see e.g. Sørensen, 1979). Usually, the windspeed reduction is not uniform over the area corresponding to the rotor swept disc. The incident wind profile, i.e. windspeed as function of height, is also not uniform, but may approximately be described as a logarithmic or power law

Fig. 4.5 Visual impact of wind converter array shown in Fig. 4.4 (Birk Nielsen, 1980)

profile. Measurements have been performed in a number of cases, on model wind turbines in a wind tunnel (in order to achieve incident wind conditions independent of time) and on actual machines. Fig. 4.6 gives an example of the measured vertical profile, before and at various stages after the passage through the turbine area. The data are from a wind tunnel experiment (Alfredsson *et al.*, 1980). It is observed that initially there is no modification of windspeed near the ground, but that after a while, during the restoration of the profile around hub height, energy is transferred into the region behind the rotor from below as well as from above. Still, the decrease in windspeed, 4–8 rotor diameters behind the

Fig. 4.6 Wind Profile Modification by Wind Turbine, based on Wind Tunnel Measurements. Abscissa: Wind speed relative to the one at top of wind tunnel. Ordinate: Height relative to hub height, taken equal to rotor diameter. Curve *a* is profile upwind from rotor, curves *b, c* and *d* are profiles 2, 4 and 8 rotor diameters downwind from rotor (based on Alfredsson *et al.*, 1980)

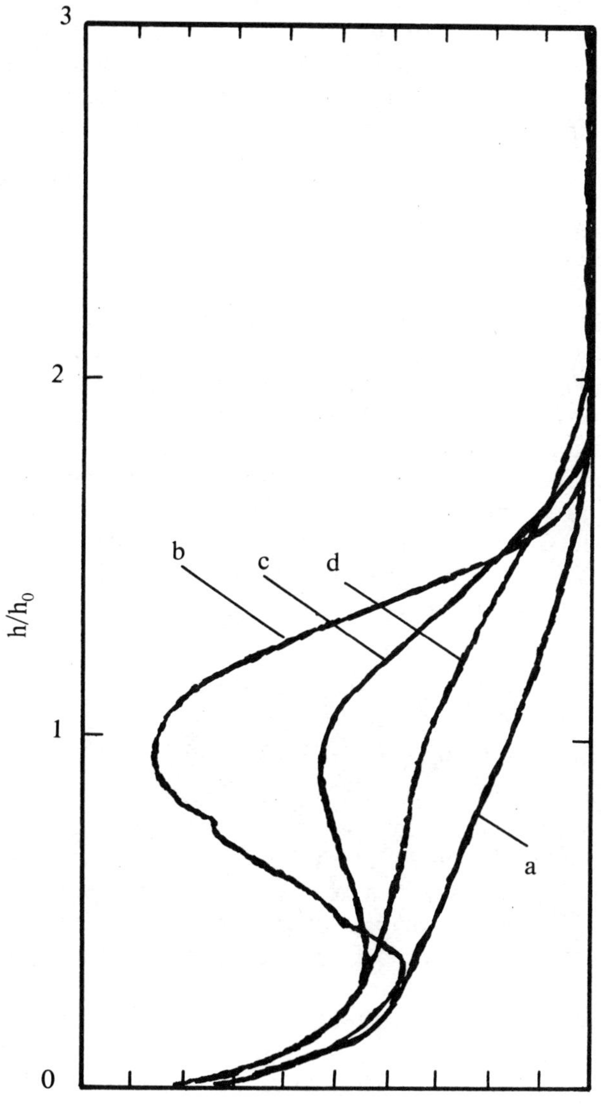

turbine is very modest near the ground. This may be taken as an indication, that changes in microclimate which is sensitive mostly to conditions very close to the ground, will be very small. This has further been confirmed by actual measurements at the US Mod-O (Rogers *et al.*, 1977). Within 0·75 rotor diameter of the machine, a 5% rainfall deficit has been deduced, in cases where the rotor is not operating. The deficit decreases and almost disappears when the rotor is turning. No significant variations were observed in air temperature, concentration of carbon dioxide, and the reduction in windspeed had generally disappeared 2–3 rotor diameters away. It is expected that this distance would be smaller for field measurements than for wind tunnel experiments, due to the natural variability of the free wind, which tends to make it impossible to distinguish wake effects from the stochastic fluctuations in the natural wind regime.

If one extrapolates the above results, the indication would be that even for extended arrays of wind turbines spaced perhaps as close as 5–10 rotor diameters apart, there would be no significant microclimatic changes, and use of the land areas between the wind energy converters, e.g. for agricultural purposes, would not be affected by the presence of the wind converter arrays.

This conclusion is likely to hold only for simple free-stream rotors. There have been proposals of a number of "advanced" wind energy conversion systems, which are all designed to concentrate the wind energy flux from a large area into a smaller area, where it can be utilized e.g. by a turbine, although this is not the only method proposed. The concentration of wind power and potential for a large energy extraction in a small region may change the conclusions concerning microclimatic impacts. Should any of these advanced concepts pass from the research stage into a stage of possible practical utilization, a specific study of the microclimatic impacts would be called for.

GLOBAL CLIMATIC IMPACTS

Wind energy utilization would have to be on a truly large scale before global climatic impacts could be imagined. This is because the formation and dissipation of wind energy is part of the solar energy cycle and is driven by the very strong forces associated with the differential solar heating of the atmosphere and the induced physical requirement for poleward transport of energy. However, the fraction of solar energy needed to maintain the general circulation in the atmosphere is very small. It is estimated to average $1{\cdot}2 \times 10^{15}$W (see e.g. Sørensen, 1980b) or 0·7% of the solar radiation at the top of the atmosphere. This equals the amount of kinetic energy in air motion lost by friction, and in an equilibrium situation also the amount of new kinetic energy created on the basis of "available energy", i.e. potential and internal (heat) energy that can be converted into motion as a result of pressure and temperature differences in the atmosphere. It also equals the rate of formation of new "available energy" by absorption of solar radiation. The equilibrium value to total kinetic energy in the atmosphere is in this model estimated at $7 \cdot 7 \times 10^{20}$J, while the equilibrium value of "available energy" is $2 \cdot 8 \times 10^{21}$J (Lorenz, 1967).

Energy extraction by wind conversion devices may be described as an increase in the frictional loss of kinetic energy. The total human energy conversion today is over two orders of magnitude lower than the $1 \cdot 2 \times 10^{15}$W frictional loss, so the wind energy utilization really has to be large in order to significantly increase this number. In this case a new equilibrium may establish itself, with the transfer rates equal to the elevated frictional loss. The equilibrium values of total kinetic and available energy will then change, but a detailed atmospheric model is required for estimating the amount of change, taking into account changes in horizontal as well as vertical energy transport in the atmosphere, as function of geographical position and altitude.

CONCLUSIONS

Most of the environmental impacts of wind energy conversion at a level which could be reached in the foreseeable future can be meaningfully assessed. Considering first the converter alone, it is evident from the short survey made above, that the impact in most areas would be considerably less than that of current energy systems, compared on a basis of equal energy production. The impacts that do exist are also generally of a different nature than the pollution and radioactivity risks associated with conventional alternatives. Generally the impacts of wind energy conversion are not associated with long latent periods or small probability-large consequence events.

It is true that the experience, particularly with large-size units, is only a few converter-years. This may have special relevance for the accident risk, where mishaps associated with prototype units may have to happen before adequate regulative standards can be enforced. Again, at this stage the consequences of such accidents are modest, and no long-range effects are involved.

With respect to the environmental impacts of transmission and energy storage systems, these are common to all energy systems, but are of course often proportional to the amount of storage and transmission needed for each system. Wind energy systems are likely to require slightly more transmission capacity than current systems, but only for massive penetration of wind in the energy systems will substantial amounts of storage be required (Sørensen, 1979). In several areas the storage needed for large-scale utilization of wind energy can be provided by hydro power systems already available. (Sørensen, 1980a). In this case no additional environmental impacts are introduced, but of course the burden of relieving such impacts, i.e. the costs, should be shared by the wind and hydro systems.

References

Alfredsson, P., *et al.,* (1980): In "Third Int. Symp. on Wind Energy Systems", BHRA Fluid Engineering, Cranfield, UK, pp. 469–484.

Birk Nielsen, F. (1980): Om Vindmollers Placering. Report from School of Architecture, Div. Landscaping and Garden Arts, Aarhus, Denmark.

Bridson, D., and Worthington, P. (1980): In "Third Int. Symp. on Wind Energy Systems", BHRA Fluid Engineering, Cranfield, UK, pp. 307–321.

Danish Dept. of Energy (1978): Newsletter No. 7 from Wind Energy Program (in Danish).

Danish Dept. of the Environment (1974): Recommendation No. 2 from Miljøstyrelsen (in Danish).

Friedman, P. (1978): In "Third Wind Energy Workshop", US Dept. of Energy, Report CONF–770921/2, pp. 517–527.

Keast, D. (1978): Noise-control needs in the developing energy technologies", US Dept. of Energy Report COO–4389–1.

Kottapalli, S., *et al.,* (1978): In "Second Int. Symp. on Wind Energy Systems", BHRA Fluid Engineering, Cranfield, UK, pp. C4–49 to 66.

Lorenz, E. (1967): The nature and theory of the general circulation of the atmosphere. Publ. No. 218TP115, W.M.D., Geneva.

Raab, A. (1980): In "Third Int. Symp. on Wind Energy Systems", BHRA Fluid Engineering, Cranfield, UK, pp. 169–182.

Rogers, S., *et al.,* (1977): Environmental studies related to the operation of wind energy conversion system. US Dept. of Energy Report No. COO–0092–77/2.

Sengupta, D. and Senior, T. (1978): Electromagnetic interference by wind turbine generators. US Dept. of Energy Report TID–28828.

Sørensen, B. (1979): Renewable Energy. Academic Press, London.

Sørensen, B. (1980a): In "Third Int. Symp. on Wind Energy Systems", BHRA Fluid Engineering, Cranfield, UK, pp. 533–543.

Sørensen, B. (1980b): Wind energy — an overview. In "Proc. UNITAR Conf. on Long-Term Energy Resources", (in press).

Spera, D. (1978): In "Third Wind Energy Workshop", US Dept. of Energy, Report CONF–770921/2, pp. 502–516.

UNEP (1981): The Environmental Impacts of Production and Use of Energy. A study by the United Nations Environment Programme. Tycooly International Publishing Ltd., Dublin.

CHAPTER 5

Energy from the Sea

M. I. SAID AHMAD
National Research Centre, Dokki, Cairo, Egypt
and
ESSAM EL-HINNAWI
United Nations Environment Programme, Nairobi, Kenya

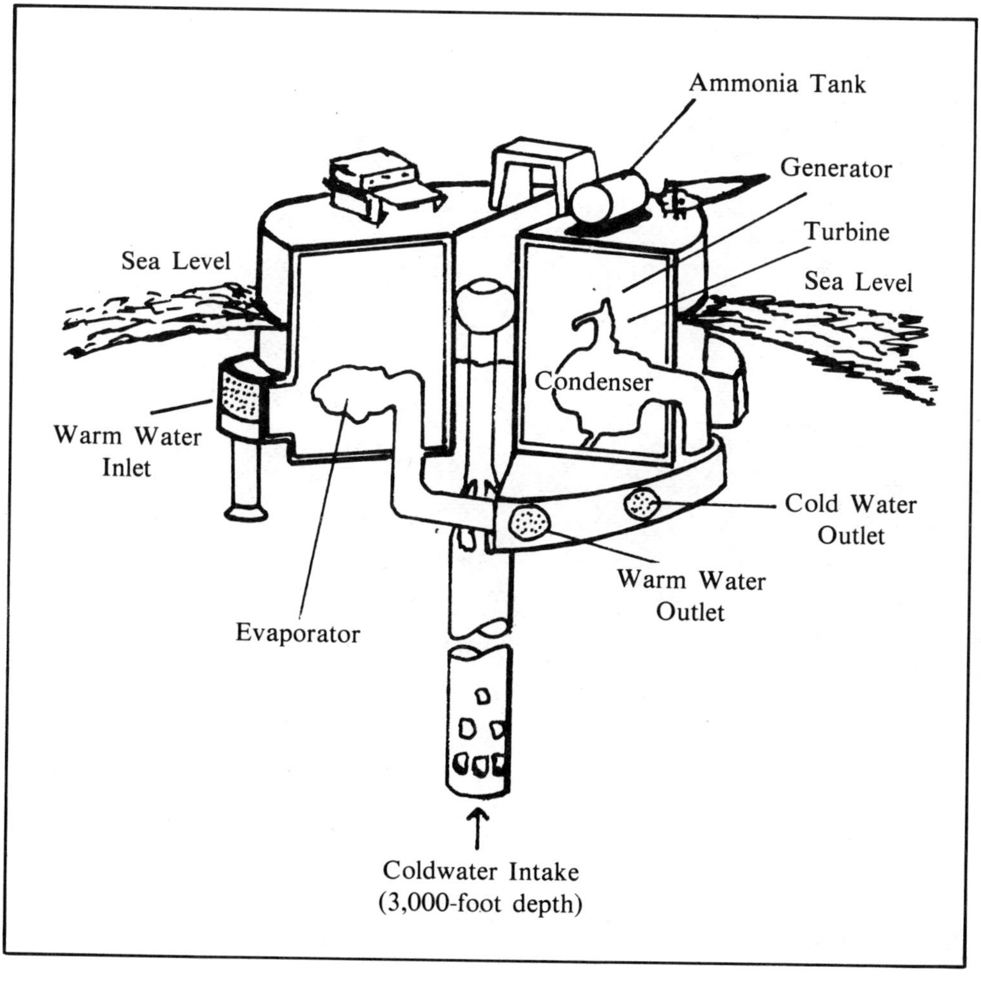

Fig. 5 OTEC Power Plant

WAVE POWER

Ocean waves constitute a renewable source. It is renewable in terms of the fairly constant conversion of solar energy into water waves through the thermodynamic processes between the atmosphere and the ocean at the air-sea interface. Latent wave power is diffuse. However, if sufficient wave power (energy flux) can be found and economically utilized, it has the potential of being a major source of energy for future use.

The energy and power in waves can be considerable. A typical mid-ocean has approximately 1.5m waves at periods of 8 seconds (Isaacs *et al.,* 1976). These waves correspond to a mean flux of wave power across a section of the ocean of the order of 10 kW/m. Summed over the entire ice-free ocean, the total power available would be 2.7×10^{12} W; the amount of power that could be extracted from this would, however, be a fraction of this amount (Merriam, 1978). Panicker (1976) estimated the total surface wave energy in the seas of the world at 12 noon, 2 October 1975 to be $1,586.3 \times 10^{15}$ joules. The corresponding total wave power estimate is 89.3×10^{15} W, which is the theoretically available power for conversion. The rate of renewal of wave power is estimated to be 10^{12} to 10^{13} W.

Waves, as such, are a low-grade form of energy and in their natural state cannot be economically exploited. Also, waves are mostly aperiodic. The low pressure head represented by a 1.5m wave would yield little power were it converted as in a hydroelectric dam of equivalent height. This is due mostly to the inherent inefficiency of hydraulic turbines operating at low pressures. Some of the design criteria for a wave powered machine may be summarized as follows: (1) the machine must have the ability to amplify the relatively low water head associated with wave height; (2) it must have a broad band range of response over wave amplitude and frequency, i.e. be non-resonant; and (3) it must have the capacity to respond to small waves as well as to withstand potentially destructive accelerations from large storm waves.

The harnessing of wave power is not a recent concept; hundreds of different schemes have been suggested since the early 1900s (see Isaacs *et al.,* 1976). However, the only device that is actually producing electricity from waves is that invented by Masuda in Japan (Masuda, 1971) and utilizes the wave motion to compress air for driving an air turbine on a free moving float. It only generates a few watts continuously but is quite suitable to power the beacon on a buoy. Units delivering a few hundred watts have been tested. The same principle is also being used to power lighthouse beacons.

Methods of extracting the energy in waves fall into three general categories: (1) utilizing the vertical rise and fall of successive waves to activate either a water or an air-operated wave-powered turbine; (2) utilizing the to and fro or rolling motion by means of vanes or cams that would turn turbines; and (3) other methods, such as concentrating water in a converging channel in which the momentum transport of shoaling and breaking waves maintains a head of water that can drive a turbine.

Devices for harnessing wave energy may be classified into propulsion schemes, buoy power supply devices, offshore power plants and shore-based power plants (see Panicker, 1976; Merriam, 1978; Count, 1979; Voss, 1979; Clarke, 1979 and Simeons,1980 for detailed description of different devices).

The engineering challenge for harnessing wave energy in power plants of several hundred kilo — and megawatts capacities is to find a cost-effective way to convert a fraction of the large amount of energy contained in wave motion under the inhospitable conditions existing in the open sea. Electric energy extracted from wave power by schemes known today might cost, according to present estimates, about 10 to 20 times the cost of electric energy from conventional generating plants (Count, 1979). World-wide application of wave energy conversion systems will be limited by geographic factors and the economics of energy transport and distribution over large distances. It is difficult to estimate the potential contribution of this source, but because the resource base is about 10 times less than geothermal and 400 times less than wind (Sørensen, 1980), it could never become a major source of power except for few favourable locations.

Environmental Impacts of Harnessing Wave Power

Because there are no large-scale wave power stations existing today it is difficult to assess the environmental effects of harnessing this energy source. Wave power plants will produce no thermal discharges or emmissions or cause changes in water salinity or require fresh water for operation. The most direct environmental impact of wave power plants is to calm the sea, since they will act as efficient wave breakers. This has beneficial effects in several locations near harbours offering safe anchorage in times of storms, and/or protecting shorelines from erosion (Nath and William, 1976; Zeigler *et al.*, 1976). However, the calming of the sea might have adverse biological effects because of the absence of waves and associated mixing of the upper layers of the sea. On the other hand, wave energy plants are likely to enhance the growth of many marine organisms by providing a more protected habitat. The obvious success of artificial reefs in attracting marine life seems to pose two problems for energy collecting devices: (*a*) Marine growths may foul machinery or collectors, causing reduced efficiency or actual destruction during storm, and (*b*) anglers and spearfishers may haunt these fishing "hot spots" in great numbers (Zeigler *et al.*, 1976). Large scale wave energy systems may influence climate, because they will interfere with the wind-driven oceanic circulation and with sea-atmosphere transfer processes.

TIDAL POWER

Tidal power is derived from the combined kinetic and potential energy of the earth – moon – sun system. Early records indicate that tide mills were installed along the Atlantic Coast of Europe, notably in Great Britain, France and Spain in the 11th century. Early tide mills were designed to extract a relatively small portion of the total energy potential producing small amounts of mechanical energy, about 30 to 100 kW, used at the immediate site (Lawton, 1972). Table 5.1 lists the operational and potential tidal power projects in the world. Suitable tidal sites have also been reported in India, the People's Republic of China, Brazil, North Korea, South Korea and Mauritius.

Table 5.1 Tidal Power Sites

Location	Average Tidal Range (m)	Average output (10^6 kWh(e)/y)	Capacity (MW)
I. *Operational*			
La Rance, France	8.5	540	240
Kislaya, U.S.S.R.	2.5	~7	0.4
II. *Projects*			
U.S.A.			
Cook Inlet, Alaska	8	10,950	2,600
Passamaquoddy, Maine	5.5	2,100	1,000
Canada			
Bay of Fundy:			
Minas Basin	11.5	10,374	3,200
Shepody Bay	9.5	2,967	920
Cumberland Basin	10	2,352	795
England			
Severn Estuary	9	20,000	4,000
Solway Firth	5	13,000	5,000
Morecambe Bay	6	10,000	4,000
Carmarthan	5.5	7,000	2,000
France			
Minquiers	8	50,000	15,000
Chausey	8	34,000	6000–12000
Argentine			
San Jose	6	9,000	1,000
Santa Cruz	7.5	4,000	
Puento Gallegos	7.5	2,000	
San Julian	6	400	
Deseado Estuary	3.5	700	
U.S.S.R.			
Mezen	6.5	2,600	1,300
Okhotsk	6		25,000
Kuloi Estuary	6.5	1,300	500
Lumbovskaya	4.5	900	400
Australia			
Secure Bay	7	1,700	600
Walcott Inlet	12	4,000	1,300

After WEC (1980), Simeons (1980)

Two physical conditions are necessary to harness the tides for power production. First, the tidal amplitude must be large – several metres at the very least. Second, the coastal topography must be such as to allow the impoundment of a substantial amount of water with a manageable volume of civil works. This means a bay with a narrow inlet, a river estuary, or a set of suitably located islands. When the coastlines of the world are examined for locations meeting these conditions only a few dozen are available; and some of these are, however, unsuitable because of inaccessibility or due to other topographical and/or environmental reasons.

Tidal power can be harnessed using the single or double basin schemes (see, for example, Lawton, 1972; Simeons, 1980). The simple single basin, single effect scheme (Fig. 5.1) has been used for centuries (though on a very small scale). The sea enters the basin through open sluices, then when the level has reached its peak the sluices are closed and the water is impounded in the basin at high tide level. When the level of the sea has fallen the water in the basin is allowed to go out through the powerhouse. Power is generated on the ebb tide only. Alternatively, the generator can be turned around and power generated on the rising tide, but this gives less energy. The civil works required are only a single dam, with movable sluice gates and a powerhouse. Another alternative of the single basin system is the single basin double effect system (e.g. at la Rance Tidal plant, France). Here the turbines used are different in design from the Kaplan turbines generally used in low head hydro-schemes. The units operate in both directions of flow during flood and ebb tides. At high tides the gates are closed. Once the necessary difference in water level has been reached between the basin and the sea. the turbines are set in motion as the flow begins from basin to sea. When the head differential falls to a level too small to allow generation, the gates are opened to reduce the level as much as possible so that once the tide turns, all the gates can be closed to keep the basin at low level until generation can restart. The turbines are set to operate in the reverse direction, until just after high tide when the cycle is once again reversed. At La Rance, France, the generating units can generate and pump in both directions. As a result, additional energy can be developed from pumping into the basin near high tide and pumping out at low tide. This additional operation increases the volume of water which can be discharged through the turbines during each tide cycle as a result of increasing the differential head; higher power output results.

Single basin schemes can generate power only intermittently, but a double basin scheme can provide power continuously, or on demand, which is an advantage. The drawback is that the civil works become more extensive (and hence cost more than the single basin scheme). The simplest double basin single effect scheme incorporates two pools, and upper and a lower one (Fig. 5.2). The power house is sited between them. Filling gates are built in between the sea and the upper pool, with emptying gates direct to the sea, between the lower pool and the sea. This set up permits continuous power generation proportionate in output to the difference in heads between the pools. The direction of flow is always from the upper pool to the lower pool. The flood tide fills the former daily while the

Fig. 5.1 Single Basin, Single Effect, Tidal Power Scheme

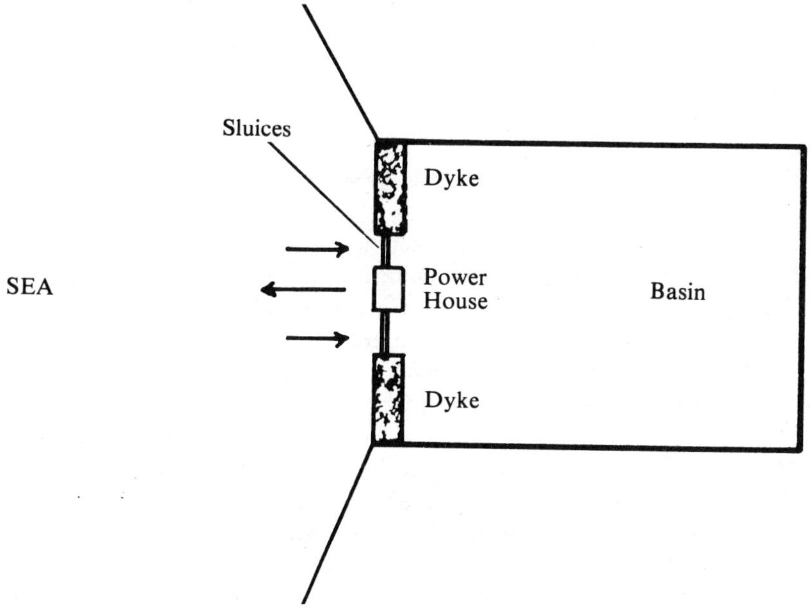

Fig. 5.2 Double Basin, Single Effect, Tidal Power Scheme

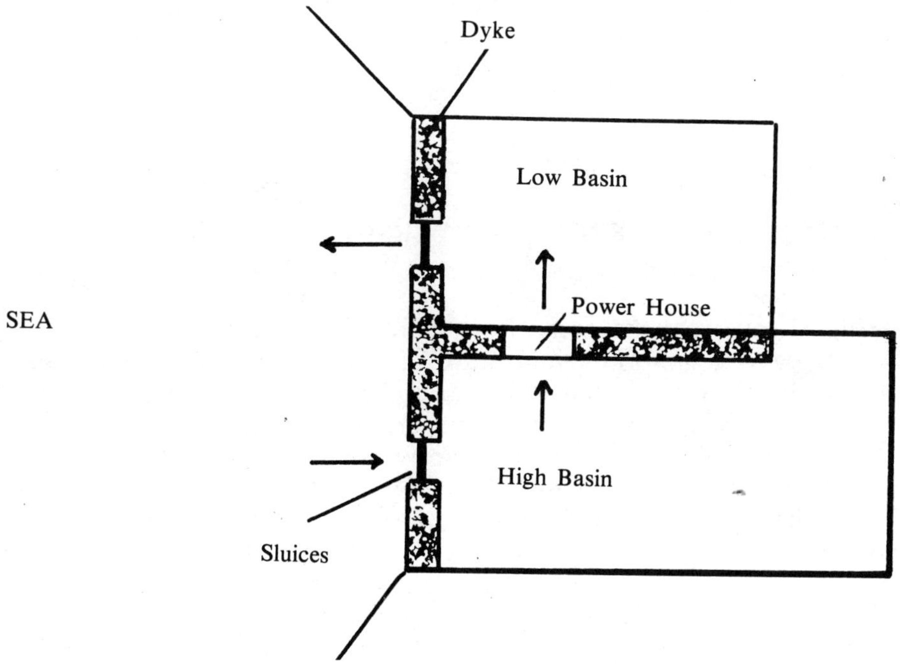

latter is emptied by the action of the ebb tide when emptying gates are opened. An example of this type of scheme is the proposed Passamaquoddy project in the U.S.A. Double basin double effect schemes involve greater construction costs not necessarily accompanied by higher output which makes them less attractive from an economic point of view.

Environmental Impacts of Tidal Power Development

The construction of a tidal power plant sets in motion a number of physico-chemical and socio-economic environmental impacts (see, for example, Waller, 1972; Shaw, 1974, 1975; Simeons, 1980). Generally speaking such impacts include inter alia, the physical obstruction created by the dams, effects on the tidal regime of the area and associated impacts on navigation and fisheries and ecological effects in the pool areas (e.g. stratification, erosion, sitlation, etc.). In some areas where waste waters are discharged from land-based sources into the sea, the building of tidal power plants would lead to serious pollution problems in the impoundments. A pollutant that, in the absence of an impoundment, was removed without nuisance by dilution and tidal exchange, might well become a problem if its concentration was higher and less dissolved oxygen was available to meet the demands of decomposing micro-organisms. The problem would be worse if settleable material was deposited, because of low velocities, in stratified deep water with an initially low dissolved oxygen concentration. Benthic organisms could be smothered and the water made intolerable for fish. It should be noted, however, that the ecological effects associated with the development of tidal power are site specific and should be assessed for each site independently to weigh the benefits against the costs of such projects.

OCEAN THERMAL ENERGY CONVERSION

The upper layers of the ocean (to about 100m depth), especially in the tropical belt, have a uniform temperature of about 28°C. The temperature decreases with depth, and at about 1000m it reaches 5°C. This thermal difference ($\Delta T°C = 23$) could be used as a renewable source of energy. The ocean thermal energy conversion (OTEC) is not a new concept. It dates back to 1882 when the French scientist D'Arsonval recognized that, by using the Carnot cycle, it was possible to utilize differences in fluid temperature to generate power. The first pilot experiment was built by George Claude at Matanzas Bay, Cuba, in 1930. The plant was to use warm sea water as the working fluid, but because of technical and economic difficulties Claude abandoned his experiment. Recently, several OTEC plants have been designed and some tested on an experimental basis (Avery, 1978; Marchand, 1979; Lavi, 1980; Kamogawa, 1980; Coleman, 1980).

The most important location of the sea thermal resource is throughout the equatorial belt from latitudes 30°S to 30°N, where deep sea water may be from 15°C to 25°C colder than surface water. However, prospective sites cannot be judged on this basis alone. Coastal characteristics can enhance or degrade the ΔT° while local costs for energy, particularly in remote islands, may be so great that OTEC can compete with conventional sources even at a relatively high unit cost. It has been estimated that by placing OTEC facilities approximately 15 km apart throughout the oceans between 20°S and 20°N latitude, the theoretical maximum limit for electricity production is of the order of 50×10^{12} W. An ultimate installed capacity representing a small fraction of this potential would still represent a significant contribution, especially, to the energy needs of islands and isolated coastal regions (Auer *et al.,* 1978; Lavi and Lavi, 1979).

OTEC POWER SYSTEMS

Two OTEC power systems have been designed: the Open Cycle and the Closed Cycle power systems (see, for example, Coleman, 1980). Both are briefly described herein.

Open Cycle System

In this system (Fig. 5.3) warm sea-water is drawn into a flash evaporator where steam is produced by flashing the sea-water. The latent heat of vaporization is provided by the 27°C sea-water as it cools by about 4°C, thus requiring the cycle to handle about 175 kg of sea-water per kg of steam produced. The sea-water effluent (sea-water cooled to about 23°C) is pumped back into the ocean at a depth sufficient to prevent it from mixing with the warm sea-water intake. The steam and entrained liquid droplets from the evaporator pass through a demister to separate the liquid droplets and any salts from the saturated vapour. The steam is then expanded through a single stage steam turbine-generator. The steam coming out of the turbine may be condensed by one of three methods (Coleman, 1980): (1) through a surface condenser that produces clean (desalted) potable water supply as condensate; (2) through a direct contact spray condenser; or (3) through a direct contact film or a sheet condenser. The last two options do not allow for production of potable water as condensate. The cooling of the condenser is achieved by cold sea-water (~5°C) pumped through the condenser from the ocean depths. The open cycle power system must eliminate the noncondensible gases that are in the warm surface sea-water. It is assumed that these gases travel with the steam through the cycle on to the condenser. These noncondensible gases could hinder the condensation process and must be driven off by a venting mechanism.

Fig. 5.3 OTEC Open Cycle Power System

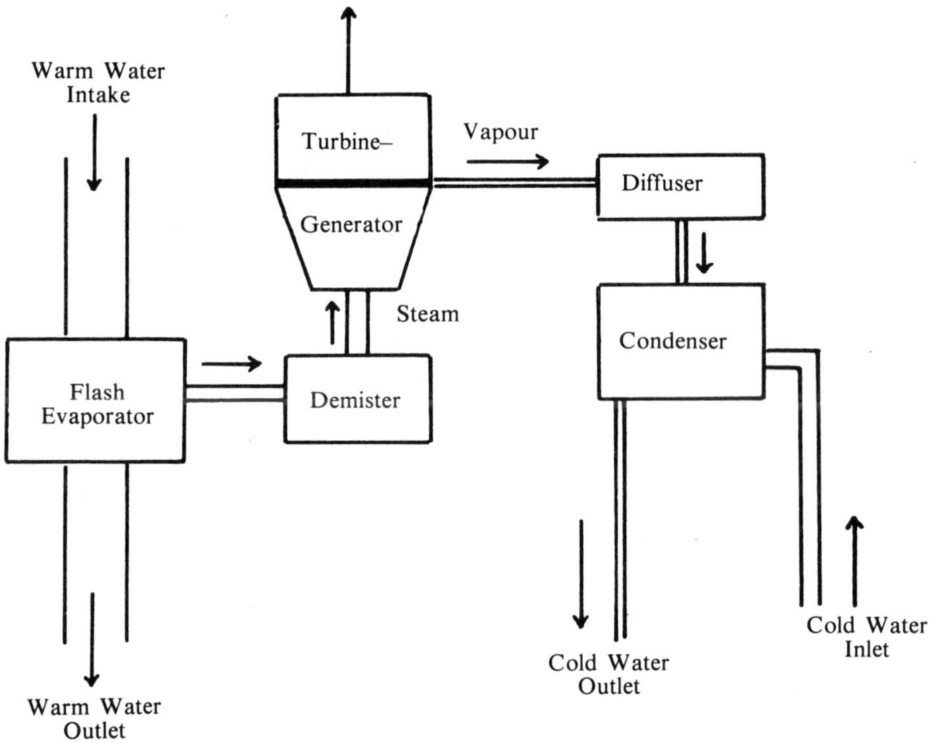

Closed Cycle System

In this system (Fig. 5.4) warm sea-water enters the OTEC plant at location 1, is pumped through the heat exchanger at location 2, and leaves the plant at location 3. The heat exchanger-evaporator (2) vaporizes the working fluid (ammonia, propane or Freon). The vapor is then expanded in a turbine (4) and leaves to enter the condenser (5). Cold sea-water enters at location 6 and flows through the heat exchanger-condenser (5). The condensed working fluid is pumped to the heat exchanger-evaporator (2) to repeat the cycle.

The electricity produced by the OTEC plant is to be: (1) transmitted to shore by cable; (2) converted into other forms of energy (e.g. hydrogen) and then piped or shipped to shore; or (3) consumed in situ by energy-intensive industry (for example, aluminium production, ammonia production, uranium enrichment, extraction of minerals, etc.). Kamogawa (1980) mentioned a case study undertaken in Japan to extract uranium from sea-water using a 100 MW OTEC plant.

Fig. 5.4 OTEC Closed Cycle Power System

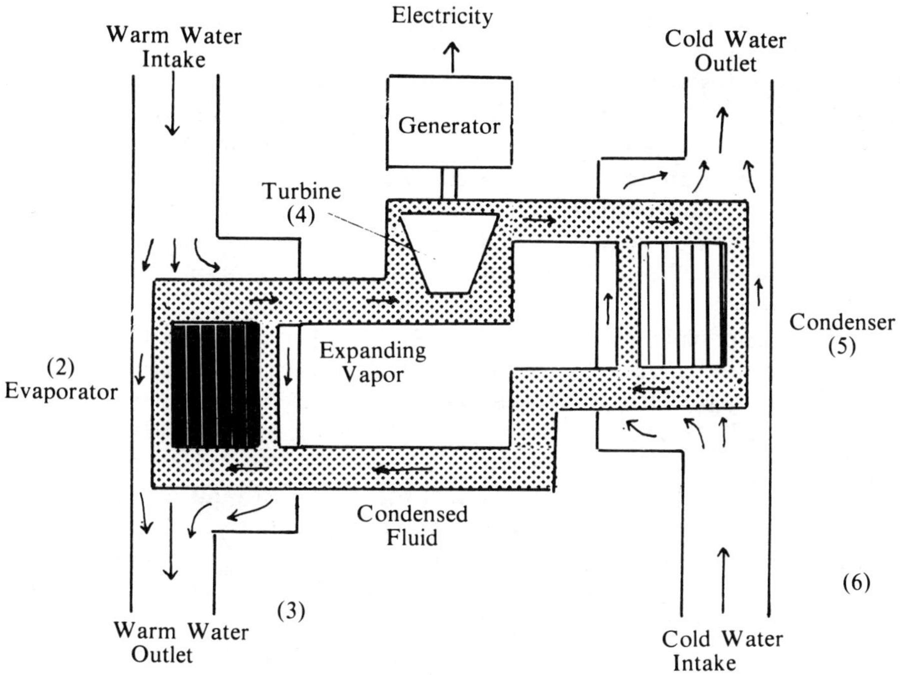

A large number of technical and economic problems are, however, associated with the development of OTEC systems. Some could be resolved through further research and development. These include improving the efficiency (the realizable overall system efficiency is at present about 2.2% which is very low compared to fossil and nuclear power plants (30–40%); see Lavi, 1980) and lowering the costs of the heat exchangers, finding a solution to potential environmental hazards created by the large amounts of ammonia used in the system, and examining alternative designes for the outer hull of the plant structure. However, other problems are inherent in the system itself (essentially a tremendously large system – one consisting of pumps, heat exchangers, and other expensive equipment – would be required to generate a relatively small amount of power). Because of the hostile marine environment, significant problems of material, operation, and maintenance would be encountered. In addition the power generated would have to be economically transported in one form or another to land-based load centres (see, for example, Zener, 1976).

Environmental Impacts of OTEC

The installation of OTEC plants raises a number of legal, economic and environmental problems that should be carefully assessed before embarking on an extensive programme for the use of ocean thermal gradients for power production.

The two dominant factors concerning the siting of an OTEC plant – The ΔT° and the mode of utilization of the electricity generated – will inevitably create conflicts with other ocean space uses (e.g., navigation, fishing, offshore mining and exploration, etc.). If the site of the OTEC facility is sufficiently far offshore to put it close to the edge of the territorial sea, or close to the territorial sea (or economic zone) of another nation, legal problems may arise due to economic and/or environmental impacts (see, for example, Stein, 1977).

Irrespective of the location of an OTEC facility, a number of environmental issues may have to be resolved. These range from questions dealing with the biological and ecological aspects of antifouling agents, primarily used for ridding the OTEC equipment of biological material which could affix itself and cause fouling, to those dealing with the ecological and environmental impacts due to changes in salinity, thermal redistribution, mixing of ocean waters, and climate (Swann, 1976; Lavi and Lavi, 1979; Lavi, 1980). The construction, operation and maintenance of OTEC plants would also have several environmental impacts on coastal zones. These include physical disruption of land and air and water pollution due to construction and maintenance facilities.

Because of the huge volumes of water involved, there is a potential danger of reduced surface temperature which may have local and global weather effects. Also, such disturbances of the thermal resource may degrade the performance of OTEC plants clustered nearby. Another environmental concern is chemical discharge from the plants. Leakage of working fluid may be objectionable depending on the working fluid selected. Leakage of ammonia in small doses is tolerable but major spills may temporarily harm the ecosystem. Fortunately, should there be a major ammonia spill, the damage to the environmental is reversible. On the other hand, leakage of a halocarbon (freon) or a hydrocarbon (propane, isobutane working fluid is potentially more serious. Likewise, discharge of biocides which may be required to control macrofouling in water passages may be environmentally objectionable. Some authors (von Hippel and Williams, 1975; Lavi, 1980) have pointed out to the possibility of substantial CO_2 release into the atmosphere caused by the operation of many OTECs. Deep cold water is highly saturated with dissolved CO_2. As the water rises to the platform to enter the condenser, the decrease in hydrostatic pressure may upset the equilibrium. It has been conjectured that the pressure drop and the temperature rise will result in CO_2 release. However, Lavi and Lavi (1979) pointed out that this may not be a serious issue because the condenser discharge will be below the surface. Further, the transit time of the water through the cold water pipe and the condenser is relatively short compared to the diffusion rate of dissolved CO_2. Also, some of the CO_2 will be eliminated from the solution through the formation of carbonates.

However, the problem of CO_2 release is of importance and it warrants further detailed assessment.

Among possible large-scale environmental hazards is the lowering of temperatures in the tropics and increasing temperatures in the mid-latitudes, because of new currents in the ocean and altered atmospheric circulation. Largescale utilization of ocean thermal energy will cause a substantial transfer of heat from the surface to deep layers so that, the cooled surface of the ocean will evaporate less water and radiate less energy back into space. The lowered heat flux from the ocean through evaporation and decreased radiation with the same energy input, through absorption of solar energy, will cause an imbalance. Part of this net input will cover the energy produced by the plant, while part should be dissipated outside the tropics, presumably by new currents warming the higher latitudes. If tropical OTEC power plants produce 65×10^{19} joules of energy (approximately the global energy consumption projected for the year 2000), there could be a drop of less than 1°C in the water of tropical oceans (Berkovsky, 1978).

On the other hand, some of the environmental impacts of OTEC plants appear to be beneficial. For example, fishing might be improved as a result of the recirculation of the nutrients (upwelling effects). OTEC plants could serve as stations for marine data gathering or for sea rescue operations.

References

Auer, P. L. *et al.* (1978): Unconventional Energy Resources. World Energy Resources 1985–2020. World Energy Conference. IPC Science and Technology Press, Guildford.

Avery, W. H. (1978): OTEC Status and Prospects. MTS Journal, April-May.

Berkovsky, B. (1978): Ocean Thermal Energy – Prospective Renewable Source of Power. Natural Resources Forum, 2, 337.

Clarke, F. J. P. (1979): Wave Energy Technology. UNITAR Conf. Long-Term Energy Resources. UNITAR, United Nations, New York.

Coleman, W. H. (1980): Westinghouse OTEC Power Systems. Energy, 5, 493.

Count, B. M. (1979): Exploiting Wave Power. IEEE Spect., Sept. 79, p. 42.

Isaacs, J. D. *et al.* (1976): Utilization of the Energy from Ocean Waves. Proc. Workshop Wave and Salinity Gradient Energy Conversion. U.S. ERDA Report No. COO–2946–1, paper F–1.

Kamogawa, H. (1980): OTEC Research in Japan. Energy, 5, 481.

Lavi, A. and G. H. Lavi (1979): Ocean Thermal Energy Conversion. Energy, 4, 833.

Lavi, A. (1980): OTEC; A General Introduction. Energy, 5, 469.

Lawton, F. L. (1972): Economics of Tidal Power. In: T. J. Gray and O. K. Gashus: Tidal Power, Plenum Press, New York.

Marchand, P. (1979): L'Energie Thermique des Mers. La Recherche, No. 105.

Masuda, Y. (1971): Wake-activated Generator. Trans. Inter. Colloq. on the Expositions of Oceans. Bordeau, France, March 1971. Th–V–T2–05, p. 1–17.

Merriam, M. F. (1978): Wind, Waves and Tides. Annual Review of Energy, vol. 3, 29. Annual Reviews Inc. Palo Alto, California.

Nath, J. H. and R. H. Williams (1976): Preliminary Feasibility Study for Utilization of Water Wave Energy. Proc. Workshop Wave and Salinity Gradient Energy Conversion. U.S. ERDA Report No. COO–2946–1, paper, I–1.

Panicker, N. N. (1976): Power Resources Potential of Ocean Surface Waves. Proc. Workshop Wave and Salinity Gradient Energy Conversion. U.S. ERDA Report No. COO–2946–1, paper J–1

Shaw, T. L. (1974): Tidal Energy from the Severn Estuary. Nature, 249, 730.

Shaw, T. L. (1975): Tidal Power and the Environment. New Scientist, October 1975, p. 202.

Simeons, C. (1980): Hydro-Power. Pergamon Press, Oxford.

Sørensen, B. (1980): Renewable Energy. Academic Press, London.

Stein, R. E. (1977): International Environmental Aspects. In: H. Gary Knight *et al.* (Editors): Ocean Thermal Energy Conversion. Lexington Books, Lexington, Massachusettes.

Swann, M. (1976): Power from the Sea. Environment, 18, 25.

von Hippel and R. H. Williams (1975): Solar Technologies. Bull. Atomic Sci., 31, 25.

Voss, A. (1979): Waves, Currents, Tides – Problems and Prospects. Energy, 4, 823.

Waller, D. H. (1972): Environmental Effects of Tidal Power Development. In: T. J. Gray and O. K. Gashus: Tidal Power, Plenum Press, New York.

WEC (1980): Survey of Energy Resources. 11th World Energy Conference. Munich 1980.

Zeigler, J. M. *et al.* (1976): Environmental Effects Arising from Salinity Gradient and Ocean Wave Power Generating Plants. Proc. Workshop Wave and Salinity Gradient Energy Conversion. U.S. ERDA Report No. COO–2946–1, paper D–1.

Zener, C. (1976): Solar Sea Power. Bull. Atomic Sci., Jan. 1976, p. 17.

CHAPTER 6

Hydroelectric Energy

ASIT K. BISWAS

Director, Biswas & Associates, 76 Woodstock Close, Oxford, England; President, International Society for Ecological Modelling; and Vice-President, International Water Resources Association.

Fig. 6 Construction at the Aswan High Dam (photo: UNESCO)

USE of water power is not a new phenomenon: man has used it in one form or another for at least some two thousand years. As to be expected, earlier attempts to harness energy from water were small-scale and for very specific purposes. Scientific and technological developments during the last one hundred years have changed this situation. Increasing demand for electricity and economic advantages derived due to economics of scale further contributed to the extensive development of large-scale hydroelectric power projects in all regions of the world.

During the Seventies, hydropower developments in many countries became more favourable compared to other forms of electricity generation due to several reasons. First, major increases in oil prices since October 1973 changed the economics of oil-fired power plants. Second, many countries were not self-sufficient in oil or natural gas, and constantly increasing balance of payment deficits due to import of oil and natural gas, forced such countries to seriously re-examine new hydro developments. Third, hydropower is the only renewable form of energy that can be used extensively in many countries immediately for large-scale generation of electricity; it is not necessary to wait for further technical developments. While the initial capital costs for hydro developments are high, operational and running costs are minimal. Like fossil fuels whose price is constantly increasing due to inflation and supply-control activities, the raw material for hydropower production, water, is available free of cost. This disparity will increase further in the future. Fourth, it does not raise the same type of environmental, ethical and safety problems as is currently associated with nuclear energy. Fifth, especially for many developing countries, much of the planning and technology necessary to develop hydropower is available locally and thus need not be imported at high costs from other countries. Sixth, the potential for future hydropower developments is very good in many countries. Many good dam sites are still available, and this is especially relevant for developing countries. Finally, hydropower is unique since the main development process provides other significant benefits. Dams are seldom constructed only for a single purpose like electricity generation; they also could simultaneously provide irrigation, flood control, better navigation, recreation, wildlife enhancement, and low flow augmentation. Thus, for developing countries, water resources developments could successfully contribute to the solution of two major crises, energy and food, and if the construction processes are planned carefully by using labour-intensive technology, they could provide employment for a large number of unskilled labourers — the main type of labour force unemployed — for several years (Biwas, 1980a).

According to the survey of energy resources presented at the World Energy Conference, held in September 1980 in Munich, hydroelectric generation currently accounts for approximately 23 percent of total global electricity supply. This is an average figure; for many countries hydropower is the dominant form of electricity generation. For countries like Brazil, Canada, Morocco, Norway and Sri Lanka, hydropower currently accounts for 70 percent to 100 percent of all electricity generated (Biswas, 1979). In a country like Canada, electricity generation authorities of individual provinces are often known as Ontario Hydro, Quebec Hydro or Manitoba Hydro because of their dependence on hydropower. Similarly, the monthly electricity bill in Canada is invariably called the "hydro bill".

While potential for hydropower developments has been exploited to a considerable extent in North America and Europe, including USSR, there is a vast potential in many countries of Africa, Asia and Latin America which remains to be developed. Annual hydraulic potentials for different regions of the world, as provided during the World Energy Conference of 1980, are shown in Table 6.1. It should be pointed out that during the preparation for the UN Conference of New and Renewable Sources of Energy, held at Nairobi in August 1981, it was felt the estimates provided for Asia may not include data from China. Currently theoretical potential for China is estimated at 6×10^{12} kWh, with technically usable potential of $1 \cdot 9 \times 10^{12}$ kWh. At the end of 1979, the total operating potential was $0 \cdot 05 \times 10^{12}$ kWh and potential under construction was $0 \cdot 0517 \times 10^{12}$ kWh.

Table 6.1 Annual Hydroelectric Potentials of Different Regions

Regions	Theoretical potential 10^{12} kWh	Technical usable potential 10^{12} kWh	Operating potential 10^{12} kWh	Potential under construction 10^{12} kWh	Planned potential 10^{12} kWh
Africa	10·118	3·14	0·151	0·047	0·201
America (North)	6·15	3·12	1·129	0·303	0·342
America (Latin)	5·67	3·78	0·299	0·355	0·809
Asia (excluding USSR)	16·486	5·34	0·465	0·080	0·368
Oceania	1·5	0·39	0·059	0·020	0·032
Europe	4·36	1·43	0·842	0·094	0·197
USSR	3·94	2·19	0·265	0·191	0·17 (estimated)
Total	44·28	19·39	3·207	1·090	2·12

Table 6.2 Estimated Probable Hydropower Development, 1976–2020

Country groupings	Potential energy in exajoules (1 EJ = 10^{18} J)			
	1976	1985	2000	2020
OECD countries	3·78	4·49	5·37	7·80
Centrally planned countries	0·72	1·20	2·88	8·70
Developing countries	1·17	1·97	4·49	11·80
World Total	5·67	7·66	12·74	28·30

With the present constantly changing energy price scenarios, and the fact that hydropower developments provide a good form of indigenous energy, it is highly likely that more and more countries will intensively develop their hydropower potentials. Estimated probable hydroelectric development during the period 1976 to 2020, as provided during the 1980 World Energy Conference, is shown in Table 6.2. It indicates that the total annual energy from installed hydroelectric facilities from all regions of the world are likely to increase five-fold during these forty-four years.

ENVIRONMENTAL IMPLICATIONS

Like any other form of energy development, hydropower also has environmental implications. It is important, however, to note a major distinction between hydropower and fossil fuel or nuclear power plants that generate electricity. In the case of fossil fuel or nuclear power plants, environmental impacts can be attributed exclusively as a direct consequence of energy development. For hydropower, the situation is somewhat different. As mentioned earlier, very few water development projects are constructed solely for hydroelectric generation. Thus, it can be argued that the environmental consequences of hydropower are actually due to water resources development, and should not be attributed to hydropower per se. In other words, if a dam is constructed, its environmental implications will remain very similar, irrespective of whether hydropower is or is not a reason for the development. It is not always easy, and often impossible, to ascribe the different environmental implications to specific purposes of water development, i.e. irrigation, power generation, navigation, etc.

Within this general framework, the environmental implications of hydropower will be examined in this chapter.

All development projects have environmental and economic consequences, some beneficial and others adverse. Whether these consequences are acceptable to the society as a whole are often matters of great controversy, and depend very much on the individuals concerned, their personal interests, views, motives, and biases. Hence, it is not exactly uncommon to find a situation where a new hydro development project is unacceptable to a certain segment of the society due to unwarranted social and environmental side-effects, while another segment may be lobbying hard for the same development for different reasons. Such controversies are quite common in developed countries, but in recent years they have surfaced in developing countries as well. A good current example is the Silent Valley development project in the state of Kerala, India, which contains that country's only substantial stretch of evergreen forests. The forest is 50 million years old, and has evolved over this long period to become the sanctuary of unique plant and animal life. The valley is called "silent" since for some unexplained reasons the noisy cicada insect is absent. Home of rare species like the lion-tailed monkey, the project has fuelled a major controversy between opponents and proponents, who typically ridicule the environmentalists for putting "monkeys before man". The dam, if constructed, will add 60 MW to the State's current 1000 MW capacity, irrigate 10,000 hectares of land in the backward region of Malabar, and provide employment for some 3,000 people during construction. To some extent, such controversies can be explained by an analysis of the nature of beneficiaries, which is seldom conducted for most projects. It is inevitable that any development project will benefit some citizens more than others, and frequently some citizens may have to bear additional costs, both tangible (i.e. heavier tax liabilities) and intangible (i.e. social and environmental costs). Resolution of such conflicts, arising within the framework of democratic pluralistic societies, through political processes, is, however, the norm rather than the exception (Biswas, 1978).

The social and environmental consequences of water development are many, and the resulting effects often extend much further than the planning area itself. The intersection of diverse forces are often so complex that ecologists and environmentalists are hard-pressed to predict overall effects with a high degree of confidence. For example, the present knowledge of ecosystems of man-made lakes leaves much to be desired. Thus, unless planning precedes construction by 5—10 years, several unpredictable and unforeseen situations tend to develop, some beneficial and others adverse. At the current state-of-the-art, environmentalists often find it impossible to convince engineers, economists and politicians that certain developments are unwise, or of the necessity to spend scarce resources on appropriate remedial measures because of lack of hard facts or solid scientific evidence. In addition, water development projects have been traditionally within the domain of engineers, and consequently social and environmental considerations have often been sadly neglected during the planning process. In certain cases, social and environmental scientists have been brought in only after the damage has become apparent (Biswas and Durie, 1971). As a general rule, it is more expensive to rectify damages once they have occurred: it is cheaper to take anticipatory remedial measures. Even though much lip-service is given to interdisciplinary teamwork, it is not that prevalent in most countries at present.

Since the environmental and social effects of hydro developments are many, these can be best discussed by dividing the effects on three categories of sub-systems: physical, biological and human. Table 6.3 provides a summary of the possible environmental implications of hydro developments (Biswas, 1977) for these three sub-systems.

Physical Sub-system

Hydropower development projects invariably change river and eco-system regimes, and thus the real question is not whether such developments will affect the environment, but rather how much change is acceptable to society as a whole, and what countermeasures should be taken to keep the adverse changes to a minimum, at a reasonable economic cost, within that acceptable range.

Table 6.3 Possible Environmental Implications of Hydro Development

PHYSICAL SUB-SYSTEM	
Hydrologic System	*Atmospheric System*
Water Quantity	Evaporation
Level	Micro-climate
Discharge	
Velocity	
Ground water	
Losses	*Crustal System*
Water Quality	Geology (soil, mineral content, structure)
Sediments	Earthquake
Nutrients	
Turbidity	
Salinity and alkalinity	
Temperature stratification	
BIOLOGICAL SUB-SYSTEM	
Aquatic Ecosystem	*Terrestrial Ecosystem*
Benthos	Submerged land and vegetation
Aufwuchs	Drawdown zone
Zooplankton	Zone above high water level
Phytoplankton	Failure impacts
Fish and aquatic vertebrates	Loss of animal habitat
Plants	Food chain repercussions
Disease vectors	
HUMAN SUB-SYSTEM	
Production System	*Sociocultural System*
Agriculture	Displacement of people
Fishing and hunting	Social costs
Wildlife	Political implications
Recreation	Anthropological effects
Energy	
Transportation	
Manufacturing	

Construction of any dam imposes an artificial barrier against free flow of water. Water starts accumulating in the newly created reservoir behind the dam, and thus inundates an extensive land area, which can no longer be used for other purposes like agriculture or forestry. The resultant flooding changes the water and ecological regimes of the area.

Although the main purpose of a reservoir is to store water, it starts to store sediments contained in the river water entering the reservoir as soon as the filling begins. Reduction of flow velocity increases sedimentation in the reservoir, which influences its storage capacity and thus its useful life. The amount of sediment carried by a river is a direct function of type, size and nature of its watershed and the land-use pattern thereon, and on the velocity and turbulence of the flow.

Since rate of sedimentation depends on the silt content of river water, some reservoirs need special planning and management from a sedimentation viewpoint, if the silt contents of the rivers concerned are high, and if the useful life of reservoirs needs to be prolonged. The yearly average silt contents of some of the major rivers are the following:

Yellow River (middle reaches)	— 37·6 kg/m^3
Colorado River	— 16·6 kg/m^2
Amur River	— 2·3 kg/m^3
Nile River	— 1·6 kg/m^3
Yangtze River	—0·4 kg/m^3

Rivers, like the Yellow River in China, create serious problems due to sedimentation. Each year it carries 1·2 billion tonnes of silt to the sea, and deposits 0·4 billion tonnes on its own channel bed. At this high rate of sedimentation, the river bed has now become higher than the surrounding areas due to siltation and the banks are protected by dikes, which have to be heightened every year. The build up of delta amounts to some 20 km^2 every year. Some of the tributaries of Yellow River carry extraordinary amount of silt. Thus, the average annual silt content of Jing and Wuding Rivers are 172 kg/m^3 and 136 kg/m^3 respectively.

The difficult problem of sedimentation can be best noted by considering the fate of Sanmen Gorge Dam on the Yellow River in China. When the reservoir impoundment started in September 1960, the storage capacity was $9{\cdot}62 \times 10^9$ m^3, and hydropower generating capacity was over 1×10^6 kw. By 1962, the reservoir capacity was reduced to $8{\cdot}12 \times 10^9$m^3 due to excessive sedimentation, and by 1964, the problem had become even more acute with an available reservoir capacity of only $5{\cdot}66 \times 10^9$ m^3. This means nearly $4{\cdot}0 \times 10^9$ m^3 of storage capacity was lost in only four years. Naturally hydropower generation suffered tremendously as well, and by 1964, it was generating less than quarter of its original capacity of 1960. The design, operation and management of the system had to be radically changed, initially to flush as much sediment as possible from the reservoir and then to prevent further rapid siltation. Much of the silt load comes during July to October when sluices are opened to flush sediments downstream. Most of the electricity is generated during November to June, when water depth increases to 30 m to 40 m as compared to 15 m to 20 m during July to October. Some of the turbines originally installed have now been moved to other sites, and the dam currently generates significantly much less electricity than what it could during the earliest years of its operation.

Sanmen Gorge Dam is an extreme example of siltation. The average annual storage loss in the United States is 3·5 percent for small reservoirs having capacity of less than 10,000 m^3. The storage losses decrease with the increase in storage capacity of reservoirs, and amount to only 0·16 percent for capacities greater than 10^9 m^3 (El-Hinnawi, 1981; Dendy et al., 1973). The storage capacity of the Bassano Dam on the Bow River in Alberta, Canada, was reduced from 34 $\times 10^6$ m^3 in 1911 to 11 $\times 10^6$ m^3 in 1970. For 19 reservoirs in Central Europe, having storage capacities ranging from 1·5 $\times 10^5$ m^3 to 0·23 $\times 10^6$ m^3, storage depletion was 0·51 percent per annum (Cyberski, 1973).

Silt also has been a major environmental problem for the Aswan Dam in Egypt. The dam, one of the largest in the world, was completed in 1968, and was built primarily for generating hydropower. Before the construction of the dam, large amounts of silt either used to be deposited on the Nile valley or carried all the way to the delta and the sea. These sediments are now being trapped in the reservoir created by the dam. Prior to the building of the dam, suspended matter in the River Nile, passing the Aswan, ranged between 100,000,000 tonnes/yr to 150,000,000 tonnes/yr. Observations made during the first few years after the completion of the dam indicate that the reservoir is losing about 60,000,000 m^3 of storage/yr due to siltation. At this rate, the dead storage capacity of 30 km^3 will be filled in about 500 years (Biswas, 1980).

As a result of the siltation in the reservoir, clean water is now flowing downstream of the dam causing erosion of the river bed and banks. One possibility now being considered is to construct a number of barrages to reduce the velocity and force of the clear water. These barrages can also be utilised for power generation. The other possibility is to spill the water into Toshka Depression located to the west of the lake (Hafez and Shenouda, 1978).

Another effect of the siltation in the reservoir is the erosion of the Nile Delta, some 1,000 km away. Prior to the construction of the dam, the Delta used to be built up during the flood season, with the silt carried by the river to the Mediterranean. This situation in the Delta compensated for the erosion that resulted from the winter waves of the preceding year. Without enough siltation, erosion of the Delta has become a major problem, and studies are now being carried out to find a suitable solution.

Loss of silt has further affected the productive capacity of the Nile Valley which used to get a regular deposit of sediments every year. Currently studies are being undertaken to assess the actual nutritive value of the silt, and the trace elements present therein, so that this loss can be compensated by using chemical fertilisers. Before the dam was constructed, the silt load carried by the Nile was estimated at 134,000,000 tonnes/yr (based on 30 years of available data), of which 125,000,000 tonnes was accounted for by the flood season. Only 12% of this amount, 16,000,000 tonnes, was deposited on land, and the rest used to flow with the water to the sea. After the dam was constructed, suspended sediment now being deposited on land is estimated at 4,000,000 tonnes, thus providing a short-fall of 12,000,000 tonnes/yr. It has been suggested that the loss of nitrogenous component of silt can be counter-balanced in the cultivated lands by some 13,000 tonnes of lime nitrate (Abul-Atta, 1978).

Lack of sediments downstream of the dam has further contributed to the significant reduction of planktons and organic carbons. It has, in turn, reduced the sardine, scombroid and crustacean population of the area. Loss of sardine along the Eastern Mediterranean has created economic problems for the fishermen who used to depend on the catch for their livelihood. Furthermore, there was a thriving small-scale industry on making bricks from the silt dredged from the canals. In the absence of such silts, many such industries have now resorted to using the topsoil near the canals to make bricks, thus contributing further to the loss of productive soil in the country. Egyptian researchers have now succeeded in making bricks out of sand, but it will be some time before the local industry can be persuaded to change from using topsoil to sand. On the positive side, however, lack of silt has reduced the cost of dredging canals.

Thermal stratification could be another problem, especially for deep lakes. During summer, thermal stratification occurs when warmer epilimnion and cooler hypolimnion are separated by a narrow thermocline. The stagnant water layer at the bottom of the lake loses dissolved oxygen due to the decomposition of organic matter. This means that the anaerobic biological population slowly starts to predominate, and they reduce nitrates to nitrites and ammonia. This is followed by sulphate reduction. Such anaerobic activities contribute to the formation of gases like methane and hydrogen sulphide which may affect water quality and thus interfere with water use. If the resulting water of a rather poor quality is discharged through a low-level penstock, it has to travel several kilometres downstream before reoxygenation can restore the water quality. This problem can be observed in many reservoirs of the Tennesse Valley Authority and in other warmer areas of the United States.

Construction of a reservoir naturally changes the groundwater regime of the area. Lateral seepage from the Lake Nasser has been estimated to reach 1,000 million m^3 per year. On the positive side, it has facilitated reclamation of low-lying arid lands at considerable distances from the lake (El-Hinnawi, 1981, 1980). On the negative side, it has contributed to the rising of the groundwater table, which has increased salinity of the soil. This has adversely affected agricultural production. However, soil salinisation and alkalinisation could become a major problem, if the project provides extensive irrigation. For hydropower production per se, soil deterioration is not normally a very serious problem.

Construction of dams and reservoirs inundate a large area of land. The Aswan Dam, for example, created a vast reservoir — Lake Nasser — having a shore-line length of 9,250 km, surface area of 6,216 km^2 and volume of 156·9 km^3 at 180-m elevation. It transformed 500 km of the River Nile from a riverine to lacustrine system.

In certain areas flooding could affect important historical monuments or archaeological sites. The best known case is the Aswan Dam which inundated areas that were thinly populated but rich in historical monuments. The magnificent temples of Abu Simbel and Philae (near Aswan) had to be dismantled and moved to higher and safer locations. The international undertakings to save such historical treasures were both complex and expensive. While very few dams constructed have affected such important historical monuments, many do threaten archaeological sites. What is important is to ensure competent archaeological surveys are carried out before inundation takes place. Even in an advanced country like the United States, where environmental awareness is at a much higher level than most other countries, many early rock paintings and petroglyphs have been already lost due to the construction of dams.

Another environmental impact of hydropower developments that affects land is due to the construction activities themselves. If the dams constructed are either rockfill or earthfill, large volumes of rock and earth have to be excavated. In order to reduce transportation costs, quarries and borrow pits are often located near dam sites. Large-scale excavation to obtain such materials not only leave unsightly scars on the landscape but also increase soil erosion and disturb wildlife. These impacts could be quite severe if large volumes of earth and rock are necessary and if the activities are confined within a somewhat limited space. Thus, for the construction of the LG 2 Dam of the James Bay project, which required about 150×10^6 m^3 of fill, and Wreck Cove Hydroelectric Project in Nova Scotia, Canada, where construction activity was confined to a small area, serious environmental problems were encountered.

Since dams are mostly constructed in remote areas, often new roads have to be constructed which necessitates removal of both vegetation and topsoil. The rate of soil erosion significantly increases due to such activities and also from the presence of many large quarries and borrow pits. This in turn increases the sediment load carried by rivers, most of which then get deposited in the reservoirs, and thus contribute to the reduction of their storage capacities. In certain cases, as in the Tarabela Dam in Pakistan, the situation could become quite serious, and it becomes essential to take countermeasures. Thus, it is necessary to ensure that construction activities are carried out within strict environmental protection guidelines.

The economic cost of siltation and sedimentation in rivers and reservoirs is quite substantial. For the United States alone, currently some 376 million m^3 of sediments are dredged every year from water bodies at annual costs of more than $300 million. Reduction of economic lives of man-made lakes further cost that nation $60 million annually (M. R. Biswas, 1979).

The presence of a large water body often has climatic implications, especially at the microclimatic level. As a general rule, the greater the surface area of the lake, the farther its influence can be observed. Measurements of all meteorological elements at the standard 2–m height indicate that the influence of man-made lakes decreases exponentially from the shore inland (Landsberg, 1979).

There is some evidence that man-made lakes in tropical climates decrease convective activity, and thus reduce cloud cover. In temperate regions, during the period prior to freezing, steam fog over the lake and often advective fog over shorelines are common conditions (Landsberg, 1979).

Meteorological studies carried out near Danjiangkau Reservoir in Hubei Province of China indicate that the reservoir has a moderating effect on the climate. Summer temperature around the lake has decreased by about 1°C and winter temperature has increased by about the same amount. A beneficial result of such a moderation has been the cultivation of oranges in an area where it was not economical prior to the construction of the dam. Currently oranges are grown much further to the north and closer to the reservoir. Precipitation has decreased near the reservoir but humidity has increased.

Humidities around man-made lakes are slightly higher in the area immediately surrounding the lake as compared to downwind locations. This is not surprising since larger the surface area inundated, more evaporation will take place. Thus, for the Aswan Dam, increasing the water level by 20 m, from 160 m to 180 m, more than doubled the lake surface area from 2,950 km^2 to 6,118 km^2, which in turn increased the annual evaporation from 6 km^3 to 10 km^3.

Finally, the possibility of inducing earthquakes by construction of large dams is another environmental problem that has not received adequate attention so far. The Koyna Dam disaster in the Indian peninsula in December 1967 which resulted in the death of 177 persons and considerable property damage, was due to an earthquake whose epicentre coincided with the dam itself (Biswas and Chatterjee, 1971). Several recent studies indicate that the observed seismic activity can be attributed directly to the creation of dams and storage reservoirs. Some of the tremors thus caused can reach magnitudes of up to six in the Richter scale and thus contribute to considerable damage. In general, the seismic disturbances can be traced to the existence of inactive faults and it seems likely that the effect of the added forces contributed by the dam and reservoir liberate orogenic tensions of much greater strength. Studies at the Kariba Dam sites appear to confirm this theory. It also appears that the height of the water column is a more important parameter in inducing earthquakes than the total volume of the reservoir. The seismic activity tends to become more pronounced once the depth exceeds 100 m (Rothé, 1973).

Seismic activities have now been recorded for several other dams and reservoirs. The earth movements which caused the Vaiont Dam disaster in Italy in October 1963, resulting in the death of more than 2,000 people, were preceded by several years of considerable seismic activity characterised by a clear relation between the frequency of shocks and the progress of filling of the reservoir (El-Hinnawi, 1981). Similarly a shock of magnitude of 4·3 in the Richter scale was observed during the filling of Manicougan 3 reservoir in Quebec, Canada, and 85 minor earthquakes, with a maximum of 4·7 in the Richter scale, have occurred at the Danjiangkau Reservoir in China. It was reported that a filling of a large reservoir in Tadzhikstan, Soviet Union, had to be halted due to increased seismic activity (Rich, 1977).

Biological Sub-system

Hydropower developments can affect the biological sub-system in many different ways, and the effects could be either beneficial or adverse.

Lands flooded due to such developments are often forest areas. In many instances timber has to be harvested before flooding commences. The final decision as to whether or not to harvest the timber depends not only on its value but also, more importantly, on the type of use of the reservoirs. Thus, if a reservoir is to be used for boating, the timber has to be harvested even if it could not be marketed. For the James Bay project in Canada, the general policy was to clear the areas where there could be intensive fishing and also from those parts that are visible from the roads and in the mouth of the tributaries. In cold temperatures, as prevailing in most Canadian waters, fully submerged trees could last a long time. For example, tree trunks in the Gounin Reservoir show very little deterioration after being flooded for some 55 years.

The effects of large-scale flooding on plant life was mentioned earlier. Flooding also affects animal life since it can inundate large areas of animal habitat. Hence, for large African dams like the Volta, major rescue operations had to be undertaken to catch and transport as many animals as possible to different and safer habitats. For large animals like elephants, giraffes or rhinoceroses, such operations are difficult and expensive. Similarly some attempts were made to relocate beavers affected by the James Bay project. Relocation efforts are especially important when rare or endangered species are involved. When unique and very unusual habitats are to be flooded, it may be necessary to change the design of the structures or consider construction in a different location. In certain cases, suitable alternatives may not be available and the projects need to be reconsidered or even stopped, if species concerned are unique.

With the filling of the reservoir, a riverine ecosystem gradually transforms to a lacustrine ecosystem. The new reservoir could become an important source of commercial fishery. Fish populations are often very large in new reservoirs during their early years. Fish production from the Lake Nasser increased from 764 tonnes in 1966 to 4,500 tonnes in 1969 and to 8,343 tonnes in 1972 (FAO, 1974; El-Hinnawi, 1981). Similarly fish catch increased in the section of the Volta River covered by the Volta Lake from 4,000 to 60,000 tonnes in only five years. Where there is little interest in commercial fishery, sport-fishing could become important, as in the Lake Diefenbaker in Canada, where such a recreational activity did not exist before.

The effect of dams, however, can be deleterious on anadromous (those hatched in freshwater but spending adult lives in sea, i.e. salmon) and catadromous (those hatched in sea but spending adult lives in freshwater, i.e. eel) fish species, unless special efforts are made, like the provision of fish ladders, which are expensive to construct.

Impoundments can also eliminate certain sensitive species. The construction of the Tellico Dam on the Little Tennessee River in the United States was delayed for more than two years because it would have eliminated the habitat of an endangered species, snail darter (*Percina tanasi*), which is protected under the Federal Endangered Species Act.

Generally changes in the flow regime reduces the diversity of benthos, and thus species composition. Fluctuations in flow rates are very detrimental to many species. If flow rates are constant, standing crop tends to increase.

Aquatic weeds also can be the result of hydropower development, especially in the tropics and semitropics. Serious problems of weed growth have been observed at Aswan, Kariba, Nam Pong (Thailand), and Brokopondo (Surinam). Unless adequate preventive measures are taken, growth of aquatic weed can be very fast. Thus, *Eichhornia crassipes,* commonly known as water hyacinth, covered an area of about 50 sq km in the Lake Brokopondo within the short period of February-December 1964. In little over 2 years, by April 1966, it had covered more than 50% of the surface of the reservoir, an area of about 410 sq km. Similarly, in Egypt, weeds were not a problem till 1964. However, in 1965, water hyacinth started to spread prolifically in Middle Egypt and the Delta area. By the beginning of the spring of 1975, various types of aquatic weeds, sometimes mixed with dense algae, had invaded more than 80% of all the watercourses and a great part of the Nile itself (Kamel, 1977). Experience in the Congo Basin has been somewhat similar. Between 1952, when the weeds were first observed, and 1955, they spread a distance of some 1,600 km, covering large areas of the Congo River. The problem of aquatic weeds is further aggravated by the presence of irrigation and drainage channels.

Weeds create several problems. First, water losses are greatly enhanced by their evapotranspiration process. If irrigation is practised, more water has to be released to ensure adequate water availability in the lower reaches. These two factors often account for a tremendous amount of water loss. Thus, at Aswan, it has been estimated that 2,875,000,000 m^3 of water is lost per year due to the preceding two factors alone. A better perspective can be obtained, if it is considered that at present prices, 2,875,000,000 m^3 of annual over-year storage on the upper reaches of the Nile at Aswan will cost some $216,000,000 which is not an insignificant figure.

Weeds also tend to increase the incidence of diseases like malaria and schistosomiasis by providing a favourable habitat to invertebrates like mosquitoes and aquatic snails, which are vectors and intermediate hosts for disease-causing agents. They further create problems by interfering with the operation and maintenance of hydroelectric generation and pumping stations, and by competing with fish for space and nutrients.

Control of aquatic weeds in the tropics and semitropics, especially after invasion, is a difficult and expensive task. Mechanised or manual clearing of weeds, especially in shallow waters, has been quite successful, but in deep waters it is not a very viable alternative. Weeds thus removed can be used to produce animal feed, biogas or manure. In certain countries like China, aquatic plants are specially cultivated as animal feed.

Chemical herbicides have been extensively used to control weeds. Thus, Egypt has a major program for spraying 2,4-D using motor launches and boats in the River Nile and the northern lakes (Kamel, 1977). Chemical control is not very effective for submerged weeds. In addition, herbicides often pose a major environmental hazard to aquatic organisms, deteriorate water quality and their long-term effects on aquatic ecosystem and human health are little understood.

The third type of control is biological, wherein fish, snails or aquatic grasshoppers are introduced to control weeds. There is still much to be learned on the use of biological controls. Naturally, the three control measures are not mutually exclusive; often they are used in various combinations for optimal weed control. The type of control measures to be used depends on various local situations such as the type of weed, density of infestation, depth and width of the channel, time of application, water use pattern, proximity of crop area sensitive to herbicides, availability of material from local or foreign sources, and availability of skilled manpower.

Water development schemes have often enhanced or created favourable ecological environments for parasitic and water-borne diseases such as schistosomiasis, dengue and dengue haemorrhagic fever, liver fluke infections, bancroftian filariasis, and malaria to flourish. These diseases are not new; for example, schistosomiasis was known during the Pharaonic times. However, incidence of these diseases is much higher if extensive perennial irrigation is practised. Schistosomiasis is currently endemic in over 70 countries, and affects over 200 million people. Over 200 million people are presently infected with malaria in the tropics and subtropics and another 250 million are infected with bancroftian filariasis (UNEP, 1978). Similarly plant growths around reservoirs provide a suitable habitat for tsetse fly to transmit trypanosomiasis to humans and domestic animals.

In contrast to the diseases mentioned, hydropower developments tend to reduce the incidence of onchocerciasis. The intermediate host, simulum fly, tends to breed in fast-flowing waters, which are often drowned by the construction of dams. Thus, the construction of the Volta Dam destroyed the breeding ground of simulum fly that existed upstream. However, adequate measures should be taken to ensure that new breeding places do not develop, especially in the fast-flowing waters near spillways.

Many cases can be cited in developing countries of Africa, Asia, and Latin America where society has had to pay heavy prices for hydro development schemes in terms of the overall health of the region as well as through ecological deterioration. Thus, such developments do not automatically and necessarily bring unmitigated benefits to mankind: they can, and often do, extract high costs as well. What is necessary is a determined attempt to minimise the costs and maximise the benefits of such developments on a long-term sustaining basis. This can only be done if ecological and environmental principles are explicitly considered in the overall planning process right from the very beginning.

Human Sub-system

Naturally all the effects mentioned earlier have direct or indirect impacts on human beings, but the distinction could be made that such developments at first contribute to environmental and ecological changes, which, in turn, affect the human sub-system (Biswas, 1980a). Under the present subheading, only those impacts on humans that directly stem from hydro developments will be considered.

Many of the major hydropower projects have contributed to major social dislocations by displacement of the local inhabitants. The Volta Dam in Ghana has inundated an area of about 3,275 sq miles, and the resulting lake has a shoreline of over 4,000 miles. As a result of the development, some 78,000 people and more than 170,000 domestic animals had to be evacuated from over 700 towns and villages of different sizes. Eventually, 52 new settlements were developed to house 69,149 people from 12,789 families. It was a major social problem since a large number of people coming from small villages (600 of the 700 original villages had less than 100 people, and only one had a population of over 4,000), and having different ethnic backgrounds, languages, traditions, religions, social values, and cultures, had to be resettled into only 52 locations. The complex emotional relationships between the different tribes and their lands were not properly understood. There were many who found it very hard to make a clean break with their ancestral roots, by leaving their gods, shrines and graves of ancestors. The development of a socially cohesive and integrated community, having a viable institutional infrastructure, became hard to achieve.

The economic stability of the settlers depended on agricultural products from family farming plots. Unfortunately land clearing schemes did not progress on schedule, and in some cases cleared areas were not ready for farming when the settlers arrived. The World Food Program had to step in to avoid a major catastrophe (Biswas, 1980b, 1979).

Similarly, the Kariba Dam on the Zambesi displaced approximately 57,000 Tonga tribesmen, who had to pay a major price for this progress. Technology transfer at that level was a major problem, since many of the planners were from outside Africa. The resettlement program for the Tonga tribesmen left much to be desired; not only did they suffer great cultural shocks when being thrust into communities as different from their own as theirs from Great Britain, but also it took 2 years to clear sufficient land to meet their subsistence needs. The government had to step in to avert famine and very serious hardships and, ironically, this well-intentioned step became one of the most destructive parts of the process. The food distribution centres became transmission sites for the dreaded sleeping sickness disease.

Innundation due to the Danjiangkau Reservoir in China displaced 320,000 people. The relocation cost was 200 million yuan, but this figure is somewhat misleading since overall labour costs during the Cultural Revolution were negligible. Currently plans are being made to increase the height of the Danjiangkau Dam, which would require relocation of some 250,000 additional people. With the present system and costs, the relocation cost is expected to be about 500 million yuan.

Similar types of consequences have been observed from other hydropower development projects. Approximately 100,000 people had to be relocated due to the Aswan High Dam, without sufficient planning, and the World Food Programme had to rush in famine relief for the Nubians. Other examples of lakes and populations displaced are the following: Lake Kainji in Nigeria — 42,000; Keban Dam in Turkey — 30,000; Sarafore Reservoir on the Volga — 23,000, and

Ubolratana Dam in Thailand — 30,000. In the United States, the Tennessee Valley Authority had to relocate 12,000 families.

The number of people that had to be resettled due to construction of dams in different parts of the world is shown in Table 6.4.

Table 6.4 Resettlement of People due to Various Dams

Dams and Dates Reservoirs Filled	Number relocated (approx.)	Countries involved
Aswan, 1968	120,000	Egypt and Sudan
Bkakra, 1963	36,000	India
Brokopondo, 1971	5,000	Surinam
Damodar, 1953 (4 projects—1959)	93,000	India
Gandhi Sagar	52,000	India
Kainji, 1969	42,000–50,000	Nigeria
Kariba, 1963	50,000–57,000	Zambia and Zimbabwe
Keban	30,000	Turkey
Kossou, 1971	75,000	Ivory Coast
Lam Pao	30,000	Thailand
11 Projects 1963–1971	130,000	Thailand
Nam Ngum, 1971	3,000	Laos
Nam Pong, 1965	25,000–30,000	Thailand
Nanela, 1967	90,000	Pakistan
Netzahualcóyotl, 1964	3,000	Mexico
Pa Mong (projected)	310,000–480,000	Thailand/Laos
Tarbela, 1974	86,000	Pakistan
TVA (*ca.* 20 dams) 1930s-present	60,000	U.S.A.
Upper Pampanga, 1973	14,000	Philippines
Volta, 1965	80,000–84,000	Ghana

Resettlement of population due to water development projects in many developing countries has not been a satisfactory experience. Inadequate planning, insufficient budgets, incomplete execution of plans and little appreciation of the problems of technology transfer have all contributed to the failure of plans. The fact that many of the people to be resettled were rural and illiterate, and thus had very little political power, did not help either. The direct beneficiaries of the projects were often the educated elites, who are in power, whereas the direct social costs were mostly attributable to the rural poor. Thus, in most cases, the nature of the beneficiaries, as distinct from the nature of the benefits accruing from the development projects, is seldom emphasised (Biswas, 1973). Yet, the nature of the beneficiaries, from the viewpoints of regional development or redistribution of income, is undoubtedly a major planning issue.

TRANSMISSION LINES

One environmental aspect associated with hydropower that is seldom discussed is the problems associated with transmission lines. Dams are often located in remote areas, and thus electricity generated has to be transmitted to the population centres. It should, however, be noted that the environmental problems arising due to the construction of transmission lines are somewhat similar, irrespective of whether electricity is generated through hydro, fossil fuels or nuclear power stations.

The corridors for transmission lines require commitments of large strips of land. Land areas used as rights-of-way for transmission lines are long and narrow, usually ranging from 33 to 61 m for single lines to as much as *ca.* 152 to 366 m for multiline corridors. According to the Solandt Commission (1975), one double-circuit 500-kV line requires a right-of-way width of 76 m, but for two such lines, rights-of-way will be at least 129 m wide, and may occasionally extend to more than 366 m.

The land area which is needed often depends on the voltage that is being used. In the United States, the highest transmission voltage in 1892 was only 10–kV. By the mid-1950s, transmission began at 345-kV, and by the late 1960s it had increased to 765-kV (Janes, 1977). By 1971, 22% of all the U.S. transmission was at 500-kV, and it is estimated that by 1980, 15% of transmission capability will be on 765-kV lines (Janes, 1977). Extensive research has already been done at the 1500-kV level (EPRI, 1975). Wider corridors may be necessary for higher kV lines, but on the other hand fewer corridors may suffice. Depending on length, one 765-kV line could replace two 500-kV or four 345-kV lines (US Nuclear Regulatory Commission, 1976).

While it is unanimously agreed that substantial land is used for transmission lines, estimates differ somewhat for land requirements per km of right-of-way. A booklet published jointly by the US Department of the Interior and the Department of Agriculture (1970), estimated that approximately 3,600 electric utilities in the United States operate over 4,800,000 km of overhead electric transmission lines, which present nearly 1,620,000 ha of land for right-of-way. It was forecast at that time that approximately 160,000 km of new transmission lines will be constructed on 607,500 ha of right-of-way, in each decade for the balance of this century. These figures have been basically accepted by later investigators (Hendrickson *et al.,* 1974; Kitchings *et al.,* 1974; Jackson *et al.,* 1978), even though one of the basic assumptions was that the demand for electric power will double in every decade.

The above estimate indicates an average land requirement of about 3·31 ha per km of right-of-way. This is approximately the same (30,000 m^3/km) as that estimated by Hammond *et al.* (1973) and the Economic Commission for Europe (1975). In contrast the US Nuclear Regulatory Commission (1976) provides an estimate of 6·06 to 9·4 ha per km. The estimates, however, are not as widely incompatible as they may appear at first sight. The difference lies in the width of the corridor considered: 33 m for the USDI–USDA estimate (Hendrickson *et al.*, 1974), and 61 m to 91 m by the US Nuclear Regulatory Commission (1976).

One possible way of looking at the environmental implications of high-voltage transmission lines is to compare them with other linear systems that compete for land, namely pipelines, expressways, and railways. This is shown in Table 6.5 (Solandt Commission, 1975). As noted in this Table, there are secondary uses of rights-of-way.

Table 6.5 Comparison of Features of Linear Land-uses

Features	500 kV Transmission Line	Pipeline	4-Lane Controlled Access Expressway	Railway
Right-of-way width (typical)	76–183 m	18 m	46–92 m	18–30 m
Effect as hysical barrier	Minimal except at towers (every 277 m)	Minimal	Yes	Yes
Secondary uses of right-of-way				
A. Farming	Yes	Yes	No	No
B. Forestry/Orchard	Yes (with 8 m height restriction)	No	No	No
Local noise	Not serious	None (except at pumping stations)	Considerable (often serious)	Intermittent
Atmospheric pollution	Negligible (ozone)	No	Yes	Intermittent
Visibility (at a distance)	50 m-high towers and lines, often visible at great distances	Limited and local	Limited and generally local	Limited and generally local

Extensive public hearings undertaken by the Solandt Commission (1975) in Ontario, Canada, give a good picture of public attitudes to, and perception of, electric transmission lines. In the case of a 500 kV line, a single circuit tower is about 36 m high and a tower capable of carrying two circuits is 50 m high. Such 50 m-high towers are all too highly visible under most circumstances. The Solandt Commission came to three important conclusions, so far as public preferences are concerned:

1. Where several circuits are required to achieve the necessary capacity and security, these circuits should be placed on as few towers as possible.
2. Where several lines of towers have to be used between two points, they should be kept side by side throughout the entire route, rather than made to run separately on different rights-of-way.
3. Where there is more than one line in the same right-of-way, the preference is to have towers located in lines abreast.

It goes without saying that the visual impact of transmission lines can be greatly reduced by putting them underground. Current estimates, however, indicate that, even though the present-day costs can be reduced by improving design methods, costs of putting lines underground would still be approximately 10 to 15 times those of overhead lines (US Nuclear Regulatory Commission, 1976).

It should also be emphasised that, contrary to popular belief, underground transmission lines will not eliminate land-use impacts. Even though the diameter of the cable would be only between 30·48 and 91·44 cm, the right-of-way requirements may exceed 31 m owing to the necessity of having access roads, adjacent storage facilities, adequate space to allow earth-moving equipment to operate, and a place to deposit excavated soil. Furthermore, the need to have free and ready access in order to monitor or carry out repair work could, unlike overhead transmission lines, severely restrict the use of the corridor for farming purposes. It has also been suggested that underground high-voltage lines may affect some types of surface vegetation (US Nuclear Regulatory Commission, 1976). If this should prove to be correct, possible alternative uses of such corridors could be further restricted (Biswas, 1980c).

It is important to prevent trees from growing into or falling onto transmission lines, as they can interrupt service. Thus, herbicides are often used to control utility rights-of-way. Plants exceeding 1·8 m in height are usually cut before herbicides are applied. One problem that often arises from such spraying of herbicides is the drift that eventually affects non-target foliage. Also, extensive research at Hubbard Brook, New Hampshire (Kitchings *et al.*, 1974), showed that where herbicide was applied after all trees and shrubs had been cut and left *in situ,* the cut area had an accelerated loss of nutrients, with losses of cations ranging from 3 to 20 times as high as from undisturbed systems. Moreover, increased erosion is often noticed, indicating soil losses.

Transmission lines can also have adverse impacts when crossing wetland areas. There is some evidence that behavioural modifications may occur for waterfowl, which could result in the absence of birds from an area within a few hundred metres of the transmission lines. The swaying of the lines in the wind, their reflective properties, and the humming of the lines, could explain the abnormal behaviour of the birds. Direct current transmission could also have effects on migratory birds using magnetic homing (US Nuclear Regulatory Commission, 1976).

Electric fields associated with a transmission line can produce a charge on human beings or other animals within the operative range of influence. For example, at certain locations, ground-level field strengths can approach appreciable values such as 10-kV/m for a 14 m high 765-kV line (Janes, 1977). In comparison, the earth's DC electric field is about 0·13-kV/m, and can be as high as 3-kV/m under thunder-clouds — even in the absence of local lightning. Questions have been raised on the resultant effects of this displacement of currents on biological systems, but any such effects are normally well below the values that are generally accepted as 'safe, let's go' levels.

Comprehensive reviews of existing studies on biological effects of high-voltage electric fields are available (Kaufman & Michaelson, 1974; Bridges, 1975, 1977). The former authors conclude that 'research to date has failed to provide convincing evidence that human exposure to stationary or low-frequency electric fields has any harmful biological effect'. According to Bridges (1975), 'although the great bulk of evidence suggests that there are no significant effects of electric fields encountered under extra-high-voltage lines, further research is needed'.

CONCLUSIONS

There is no doubt that the primary effects of the vast majority of hydropower development projects around the world have been beneficial. Equally, however, there is no doubt that many of these development projects have contributed to unanticipated adverse environmental effects, some of which could have been eliminated and others significantly reduced in magnitude by using appropriate planning processes. Furthermore, there appears to be a considerable difference of opinions on criteria and techniques by which successes or failures of projects can be judged. Hence, it is not exactly unusual to find a major water development project hailed as a technological triumph by engineers, welcomed as a success in terms of economic efficiency and regional income distribution by economists, but seriously questioned as to its desirability by environmentalists and sociologists. Such an anomalous situation can often be attributed to the lack of adequate interdisciplinary interactions during the planning process, which means the objective functions that are being maximised by different groups are not the same. Lack of public participation during the planning and construction phases further complicates the situation.

The addition of environmental quality in recent years to the other two traditionally accepted objectives of hydro developments — economic efficiency and regional income redistribution — has made the planning process more complex than ever before. Inclusion of environmental quality as an objective of development recognises the fact that the welfare of the society has other dimensions besides economics, and hence the real question is not whether environmental quality should be considered as a planning objective, but rather how it should be incorporated objectively within the planning framework. Existing analytical techniques available for making planning decisions cannot effectively cope with the social, ecological and environmental concerns. To give just one example, economic analyses in terms of marginal benefit-cost consideration cannot handle these types of complex problems. Damages to the environment caused by the construction of a dam, whether to the beauty of a canyon or the countryside, or to the overall ecology, cannot be analysed by the fine tuning of marginalism. Neither can this approach be successfully used where benefits are short-run and quantifiable while the costs are long-run and often unknown and unquantifiable (Biswas and Coomber, 1973). Such analyses often involve the question of choice or value judgement. The cost of upgrading the quality of water, air or land is measurable, but the benefits accrued by improving an additional unit of water, air or land quality is immeasurable, and hence in the realm of values. Such situations, however, are not unique: like many other social choices, decisions and value judgements have to be made through the political process.

Finally, planning is for the people, and planners must give adequate emphasis to the social and environmental consequences stemming from hydro development projects. Technical and economic feasibility studies are needed, but equally necessary are social and environmental feasibility studies. Without such analyses, public understanding and acceptance of the programme, an important consideration to judge the success of water development projects, may not be complete. Planners and engineers must learn from past mistakes committed, and should not make similar mistakes all over again. The planning process should become more sensitive to social and environmental problems, since long-term sustaining developments can only take place within the framework of appropriate environmental guidelines: otherwise the overall strategy of development will be self-defeating. Harmony can come only with integrated planning but discord is comparatively easy to produce.

References

Biswas, Asit K., 1980a, "Labour-based Technology for Large Irrigation Works: Problems and Prospects", Technology and Employment Programme, Working Paper 63, International Labour Office, Geneva, 37 p.

Biswas, Asit K., 1980b, "Environment and Water Development in the Third World", Journal of Water Resources Planning and Management Division, American Society of Civil Engineers, Proceeding Paper 15295, Vol. 106, No. WRI, pp. 319–332.

Biswas, Asit K., 1980c, "Non-Radiological Environmental Implications of Nuclear Energy", Environmental Conservation, Vol. 7, No. 3, pp. 229–237.

Biswas, Asit K., 1979, "Water: A Perspective on Global Issues and Politics", Journal of Water Resources Planning and Management Division, American Society of Civil Engineers, Proceeding Paper 14815, Vol. 105, No. WR2, pp. 205–222, Reprinted in Water Resources Journal, Economic and Social Commission for Asia and Pacific, United Nations, Bangkok, December 1980, pp. 30–41.

Biswas, Asit K., 1978, "Environmental Implications of Water Development in Developing Countries", Water Supply & Management, Vol. 2, No. 4, pp. 283–297.

Biswas, Asit K., 1977, "Environmental Implications of Energy Development", Engineering Issues, American Society of Civil Engineers, Proceeding Paper 12681, Vol. 103, No. EI1, pp. 49–59.

Biswas, Asit K., 1973, "Socio-economic Considerations in Water Resources Planning", Water Resources Bulletin, Vol. 9, No. 4, pp. 746–754.

Biswas, Asit K., and Chatterjee, S., 1971, "Dam Failures: An Assessment", Engineering Journal, Engineering Institute of Canada, Vol. 54, No. 3, pp. 3–8.

Biswas, Asit K., and Coomber, N.H., 1973, "Evaluation of Environmental Intangibles", Genera Press, Bronxville, New York, 77 p.

Biswas, Asit K., and Durie, R. W., 1971, "Sociological Aspects of Water Development", Water Resources Bulletin, Vol. 7, No. 6, pp. 1134–1143.

Biswas, Margaret R., 1979, "Environment and Food Production", in "Food, Climate and Man", Edited by Margaret R. Biswas & Asit K. Biswas, John Wiley & Sons, New York, pp. 125–158.

Bridges, J. E., 1977, "Environmental Considerations Concerning the Biological Effects of Power Frequency (50 or 60 Hz) Electric Fields", in Proceedings IEEE PES Winter Meeting, Paper F77-256-1, Institute of Electrical and Electronics Engineers, New York, NY, 17 p., mimeogr.

Cyberski, J., 1973, "Accumulation of Debris in Water Storage Reservoirs of Central Europe", in "Man-Made Lakes: Their Problems and Environmental Effects", Edited by W. C. Ackerman, G. F. White and E. B. Worthington, Geophysical Monograph No. 17, American Geophysical Union, Washington, D.C., 847 p.

Denby, F. E., *et al.,* 1973, "Reservoir Sedimentation Survey in the United States", in "Man-Made Lakes: Their Problems and Environmental Effects", Edited by W. C. Ackerman, G. F. White and E. B. Worthington, Geophysical Monograph No. 17, American Geophysical Union, Washington, D.C., 847 p.

Economic Commission for Europe (cited as ECE), 1975, "Environmental Aspects of Energy Production and Use, with Particular Reference to New Technologies", Report ENV/R. 43, ECE, 15 December, Geneva, Switzerland, 42 p.

Economic Commission for Europe, 1978, "Environmental Aspects of Energy Production and Use, with Particular Reference to New Technologies", Report ENV/R. 43/Add. 2, 12 January, 15 p.

Electric Power Research Institute (cited as EPRI), 1975, "Transmission Line Reference Book 345 kV and Above", Fred Weidner and Son, New York, N.Y.

El-Hinnawi E. E., 1981, "Environmental Impacts of Production and Use of Energy", Tycooly International Publishing Ltd., Dublin 322 p.

El-Hinnawi E. E., 1980, "The State of the Nile Environment", Water Supply & Management, Vol. 4, No. 1, pp. 1–11.

Food and Agricultural Organization (FAO), 1974, "Lake Nasser Area, Agricultural Potential", Technical Report 4, FI: DP/EGY/66/558, FAO, Rome.

Hafez, M., and Shenouda, W. K., 1978, "The Environmental Impacts of the Aswan High Dam", in "Water Development and Management: Proceedings of the United Nations Water Conference", Part 3, Pergamon Press, Oxford, pp. 1777–1785.

Hammond, A., Metz, W. and Maugh, T., 1973, "Energy and the Future", American Association for the Advancement of Science, Washington, D.C., 180 p.

Hendrickson, P. L. Bahl, R. W., Gray, B. A. and Maynard, W. S., 1974, "Measuring the Social Attitudes and Aesthetic and Economic Considerations which Influence Transmission-line Routing", Battelle Pacific Northwest Laboratories, Richland, Washington, Report BNWL — 1837, 7 p.

Jackson, R. H., Hudman, L. E. and England, J. L., 1978, "Assessment of the Environmental Impact of High-Voltage Transmission Lines", Journal of Environmental Management, Vol. 6, No. 2, pp. 153–170.

Janes, D. E. 1977, "Background information on High-Voltage Fields", Environmental Health Perspectives, Vol. 22, pp. 141–147.

Kamel, J., 1977, "Aquatic Weed Problems in Egypt", United Nations Water Conference, Mar del Plata, Argentina, Document No. E/CONF. 70/TP 20, 8 p.

Kaufman, G. E. and Michaelson, S. M., 1974, "Critical Review of the Biological Effects of Electric and Magnetic Fields", in "Biologic and Clinical Effects of Low-frequency Magnetic and Electric Fields", Edited by J. G. Llaurado *et al.*, Charles C. Thomas, Springfield, Illinois.

Kitchings, J. T., Shugart, H. H. and Story, J. D., 1974, "Environmental Impacts Associated with Electric Transmission Lines", Oak Ridge National Laboratory, Oak Ridge, Tennessee, Report ORNL-TM-4498, pp. 1–41.

Landsberg, H. E., 1979, "The Effects of Man's Activities on Climate", in "Food, Climate and Man", Edited by Margaret R. Biswas & Asit K. Biswas, John Wiley & Sons, New York, pp. 187–236.

Rich, V., 1977, "USSR", Nature, Vol. 265, p. 201.

Rothé, J. P., 1973, "Summary: Geophysics Report", in Man-Made Lakes: Their Problems and Environmental Effects", Edited by W. C. Ackerman, G. F. White & E. B. Worthington, Monograph No. 17, American Geo-physical Union, Washington, D.C., pp. 441–454.

Solandt Commission, 1975, "Transportation: A Public Inquiry into the Transmission of Power Between Lennox and Oshawa", Report to the Government on Ontario, Toronto, 213 p.

United Nations Conference on New and Renewable Sources of Energy, 1981, "Report of the Technical Panel on Hydropower on Its Second Session", General Assembly, A/CONF. 100/PC/30, United Nations, New York, 38 p. + annexes.

United Nations Environment Programme, 1978, "Environmental Issues in River Basin Development", in "Water Development and Management: Proceedings of the United Nations Water Conference", Part 3, Pergamon Press, Oxford, pp. 1163–1172.

US Department of the Interior and Department of Agriculture, 1970, "Environmental Criteria for Electric Transmission Systems", US Government Printing Office, Washington, D.C.

US Nuclear Regulatory Commission, 1976, "Nuclear Energy Centre Site Survey — 1975: Part III, Technical Considerations", US Nuclear Regulatory Commission, Washington, D.C.

CHAPTER 7

Energy Storage Systems

H. JACK ALLISON
National Science Foundation
Washington, D.C., U.S.A.

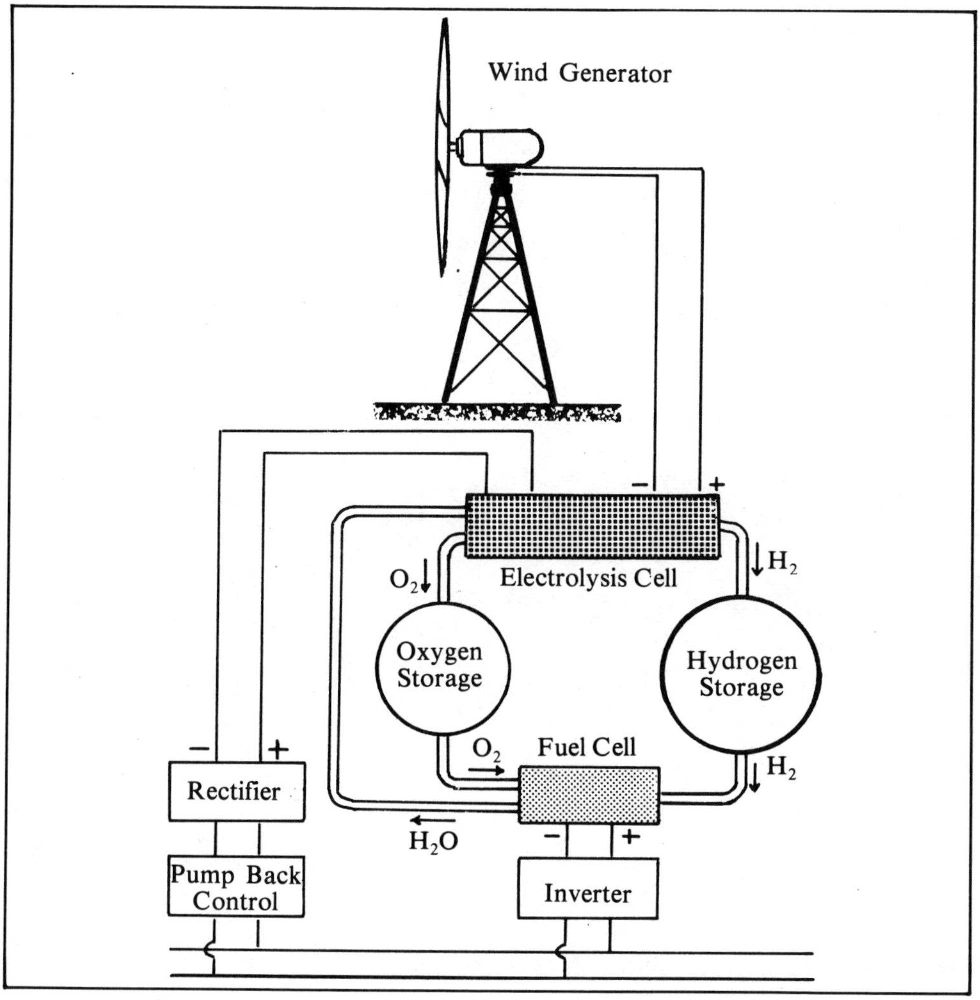

Fig. 7 Energy Storage Through Electrolysis

INTRODUCTION*

RENEWABLE energy sources are expected to make a significant contribution to the power systems of the future. Their diffuse and intermittent character, however, create problems of collection and storage which must be solved before the potential of such resources can be fully realized. A variety of systems which convert solar energy into thermal, mechanical, or electrical energy are presently under development in research institutes throughout the world. Such systems must include some mechanism of energy storage if they are to compete with conventional power systems, which can supply energy on demand. Energy storage systems can also be used in conventional power systems to store energy when demand is low, then insert the stored energy into the power grid when peak load conditions occur. (see, for example, Hughes *et al.,* 1962; Fernandes, 1975; Ramakumar, 1975, 1976; Ramakumar *et al.,* 1969, 1974, 1975; Salihi, 1973; Dickson *et al.,* 1974).

Proponents of renewable energy systems generally refer to the benign environmental character of such systems as a fundamental reason for turning to them for our energy needs in the future. Such judgments may prove to be premature, especially when energy storage subsystems are developed to increase the applicability and competitiveness of solar energy systems. This chapter addresses that conflict. It consists of a survey of the potential environmental problems associated with the energy storage systems currently being developed to serve the needs described above.

CLASSIFICATION OF ENERGY STORAGE SYSTEMS

Depending on the form in which energy is stored, energy storage systems can be classified as follows:

I Storage in mechanical form:

(a) as potential energy

(i) pumped storage of water

(ii) compressed air (direct or isothermal)

(b) as kinetic energy

(i) in flywheels.

* The views expressed in this Chapter are those of the author and not necessarily the positions or the policies of the National Science Foundation.

II. Storage in chemical form:
- (*a*) secondary batteries
- (b) hydrogen (in compressed gaseous, liquid or hydride form)
- (c) synthetic fuels
- (d) rechargeable fuel cells.

III Storage in thermal form:
- (a) in sensible heat systems
- (b) latent heat storage
- (c) thermochemical systems.

IV. Storage in electrical (electromagnetic) form:
- (a) in high intensity electric fields
- (b) in magnetic fields (superconducting magnet storage).

With the exception of pumped hydro storage, none of the above energy storage techniques have reached the advanced prototype stage required before their true potential can be documented.

MECHANICAL STORAGE

Pumped Hydro

For decades, utilities have used pumped water storage (Rogers and Allen, 1974) as an economical means of utilizing off-peak energy to pump water to a reservoir at a higher elevation. During peak load periods, the stored water is released through the pumps, now acting as turbines, to drive synchronous machines and generate electricity to meet the peak demand. On a smaller scale, farmers and ranchers have used windmills (and, more recently, solar pumps) to pump water into reservoirs for later use. These techniques require suitable terrain and some capital investment, and the environmental consequences associated with such systems are well known. They include the problems associated with terrain modification and the displacement of native vegetation and animal populations and the potential damage to surrounding areas in the event of dam failure.*

Topographic problems and conflicting land requirements have recently stimulated interest in underground pumped water storage. In this process, energy is stored during off-peak hours by pumping water from an underground reservoir

* See also Chapter 6.

to a surface reservoir, where it is stored for later use. When peaking demands occur, the water is allowed to return to the lower reservoir, driving a water turbine and a synchronous generator enroute. The round-trip efficiency for such a storage system is between 70 and 75 per cent. The upper reservoir may be an existing body of water, and the lower reservoir is created by excavating a cavern the desired distance below the surface reservoir. A schematic diagram illustrating this energy storage technique is shown in Fig. 7.1.

Fig. 7.1 An Isometric View of an Underground Pumped Storage Plant (After: Scott, 1977).

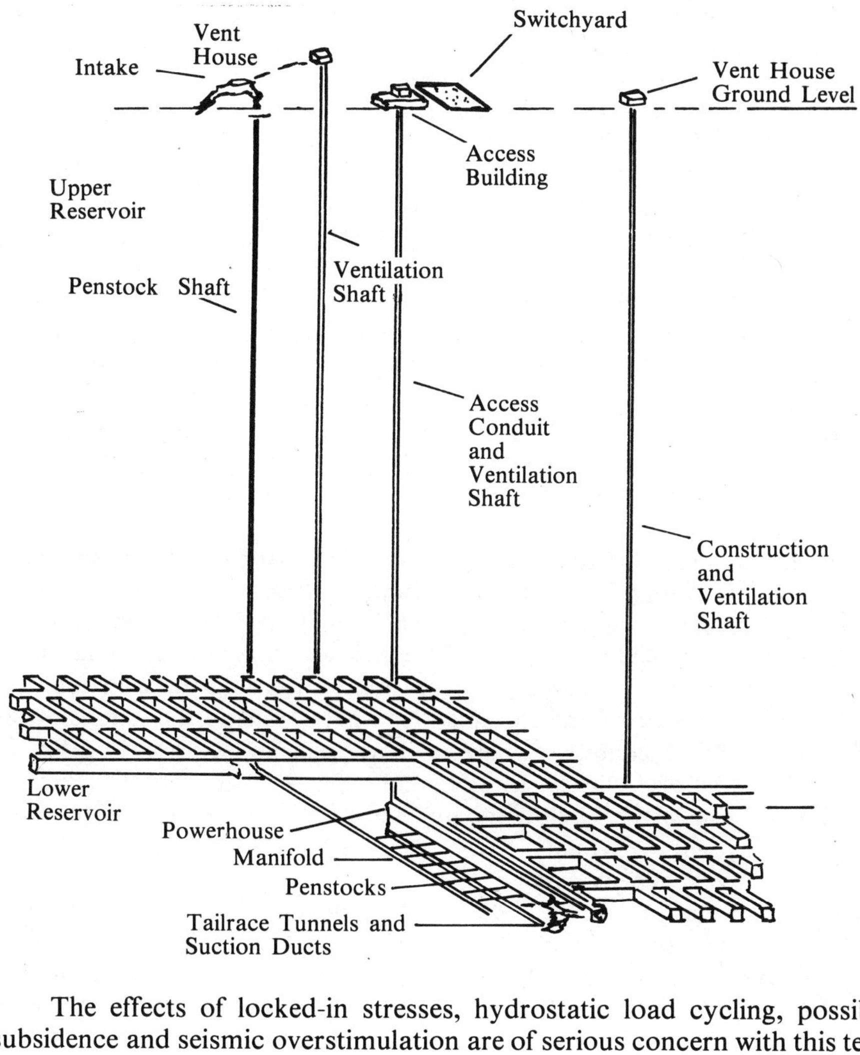

The effects of locked-in stresses, hydrostatic load cycling, possible land subsidence and seismic overstimulation are of serious concern with this technique of energy storage. The possible contamination of surface reservoirs by bacteria, leached materials and trace elements from the subsurface storage facility poses a risk to human health as well as aquatic flora and fauna. Aquatic species may also be adversely affected by the dramatic and regular changes in water levels in the surface reservoir which are an inevitable consequence of this energy storage technique.

Compressed Air

Energy storage in the form of compressed air stored in natural or man-made underground caverns (Day *et al.*, 1974; Woodbridge, 1970) represents an interesting alternative to pump-back water storage systems. The stored energy can be reutilized by using low pressure turbines or piston engines. This approach, however, results in very low (30%) round-trip efficiencies. A superior approach would be to expand the air through a combustor and turbine-generator system. This approach offers the advantage that the turbine is not required to drive its compressor, thus tripling its output power and halving the fuel consumption per kWh as compared to low pressure turbine or piston engine systems.

Combined compressed air storage-gas turbine systems can have round-trip efficiencies of 45%, and they offer potential for lowering generation costs for utility peaking power. They also require minimal above-ground land area and utilize proven and existing gas turbine technology. A schematic diagram of this energy storage approach is shown in Fig. 7.2.

Compressed air storage systems can operate either with variable pressure in the lower reservoir (dry cavity) or with hydraulic pressure compensation obtained by pumping water from a nearby reservoir or lake to further improve its operation.

In isothermal air compression (also known as hydraulic air compression or "trompe"), water at a higher elevation is allowed to flow down a vertical pipe through a circumferential venturi entrance. At the minimum cross-section of the venturi, openings are provided so that the water flowing past these openings will induce air to be sucked into them and become entrained with the water stream. With sufficient water velocity, the entrained air will be transported to the bottom of the pipe to a closed chamber where it is trapped and compressed by the conversion of its kinetic energy to potential energy. The chamber can be pressurized to a value slightly below the hydraulic head of the vertical tube. An efficiency of 82 percent is feasible for such an installation. Fig. 7.3 illustrates the operation of the hydraulic air compressor.

Salt domes are attractive candidates for compressed air storage systems. However, the disposal of solution mining brines in saline aquifers may displace waters or spillover to fresh water aquifers. In addition, surface disposal of such brine solutions in evaporation ponds may lead to leaching or runoff problems. Toxic elements may also be leached from hard-rock mine tailings and contaminate surface (or ground) water supplies.

Fig. 7.2 Compressed Air Storage System.

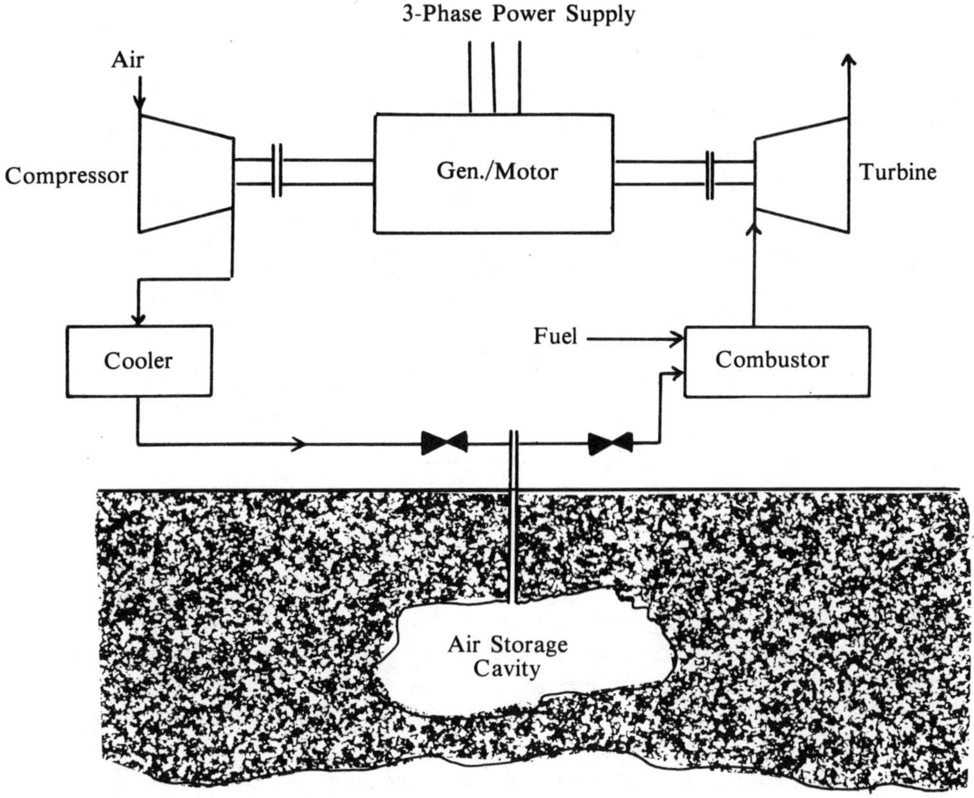

Groundwater flow patterns and hydrology characteristics may be affected when compressed air storage systems are installed in aquifers. Mineralization of aquifers may be accelerated by the increase in underground temperature caused by a compressed air storage system. Finally, the problems of land subsidence and seismic overstimulation during cycling of this storage technique must be considered in any effort to identify the potential environmental impact of compressed air storage systems.

Flywheels

Kinetic energy can be stored in spinning objects such as flywheels (Rabenhorst *et al,* 1973; Lawson, 1975; Fullman, 1975). The amount of energy stored is proportional to the square of the speed of rotation. For example, a two-to-one speed variation permits 75% energy extraction. A bilatereal transmission system is required for transferring energy in and out of the flywheel.

Fig. 7.3 An Isothermal Air Compression Energy Storage Technique.

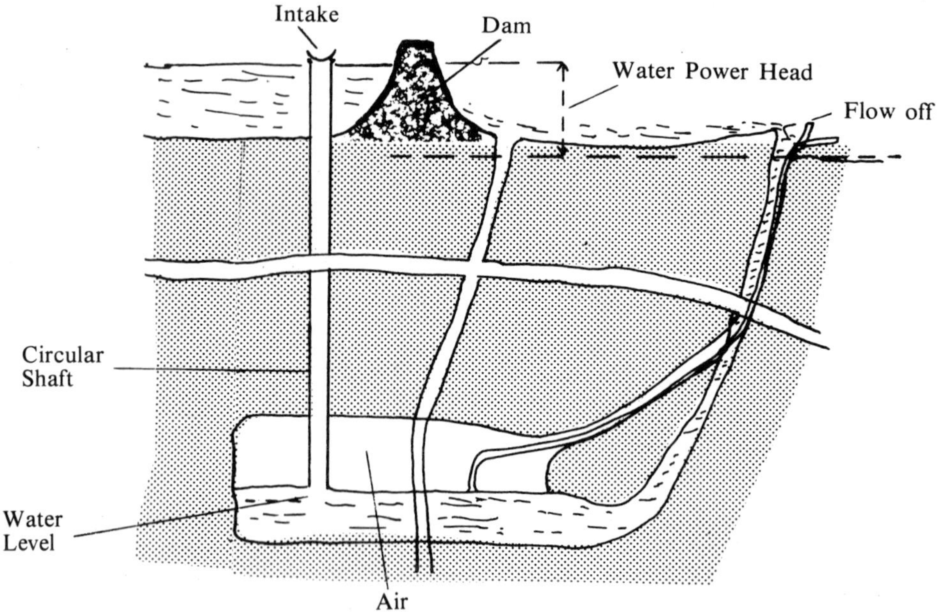

Materials suggested and used for making advanced flywheels range from special steel to various combinations of epoxy and other materials such as glass, graphite and Kavlar. Specific energy densities range from 4·4 to 15 Wh/kg and 35 to 99 kWh/m^3 for systems without containing rings for a life of 10^4 cycles. For systems with containing rings, the ranges are 4·4 to 11 Wh/kg and 17·7 to 70 kWh/m^3. Research and development work is needed to build and demonstrate the capabilities of various materials proposed and to improve the cost effectiveness of flywheel energy storage systems.

The primary objective in flywheel research and development is to provide safe and economical systems for regenerative braking in electric or hybrid vehicles. This application of flywheels has significant potential for increasing the efficiency and battery life of such vehicles. Primary environmental concerns for this energy storage technique involve the potential hazards associated with a sudden release of stored energy when the flywheel disintegrates (due to accidental or routine failure), and the possibility of toxic gas releases from fiber-epoxy materials at elevated temperatures which may result from such a system malfunction.

CHEMICAL STORAGE

Secondary Batteries

Energy can be stored in chemical form in secondary batteries, (Douglas, 1974; Yao and Birk, 1975) the most well known of which is the lead-acid battery.

In their present state of development, lead-acid batteries cannot satisfy the requirements of either electric cars or storage of off-peak electric energy. Significant improvements are necessary to achieve 20-year cell life and 5000 or more deep discharge cycles. Other commonly available batteries such as nickel-cadmium and silver-zinc batteries are presently too expensive to be used on a large scale and are severely limited by material availability. Nickel-iron batteries, though rugged, suffer from low volumetric energy density, high gas evolution, and low efficiency.

Advanced battery stems under development include the following types:

(a) High Temperature Batteries (300°–400°C)
  (i) Lithium-sulphur
  (ii) Sodium-sulphur

(b) Medium Temperature Batteries (100°C)
  (i) Sodium Chloroaluminate

(c) Low Temperature Batteries (room Temperature)
  (i) Aquenous zinc-chlorine
  (ii) Aqueous zinc-oxygen (or air)
  (iii) Aqueous zinc-nickel oxide
  (iv) Nickel-hydrogen.

Available performance data for some of the secondary batteries now under development is summarized in Table 7.1. Round-trip efficiencies of 70 to 87% are being predicted for such advanced battery systems. Initial application of battery storage systems is expected to be to electric vehicles.

The charging and operation of lead-acid batteries can result in the emission of toxic gases such as arsine (AsH_3) and stibine (SbH_3). Hydrogen is also evolved from such batteries during fast charging cycles. Other batteries can be expected to evolve a combination of toxic and explosive gases during charging, depending on the electro-chemical operations associated with their design.

Table 7.1 Characteristics of Secondary Batteries

Type	Wh/kg*	kWh/m^3*	Life Cycles
Lead-Acid (Motive Power)	11/22	49/92	1500/2000
Lead-Acid (Elec. Vehicles)	21/36	56/99	500/1000
Silver-Zinc	100/120	244/310	100/300
Nickel-Cadmium	33/40	92/127	300/2000
Nicel-Iron	22/33	32/49	3000/3000
Nickel-Zinc	66/88		250/350
Zinc-Chlorine	66/66		500/500
Sodium-Sulfur	171/222		1000/1000
Lithium-Sulfur	128/171		1000/1000

* The two numbers correspond to 1/6 hour discharge rates.

Lithium/metal sulphide batteries can be especially hazardous under malfunction or accident conditions. The lithium ion is toxic to the human central nervous system. Further, the accidental release of LiC1-KC1 electrolyte at an operating temperature in excess of 400°C, coupled with the generation of hydrogen gas during oxidation of lithium-aluminium alloy in the presence of moisture, represents a severe threat to the local environment.

Sodium-sulphur batteries may release liquid sodium and sulphur (at 300 to 350°C) into the environment when they malfunction. Sodium can react with air to emit toxic fumes of sodium oxide, and with water to generate hydrogen and heat under such conditions. Sulphur, when heated in air, can oxidize and emit toxic fumes. Similar problems exist with zinc-cholorine or zinc-bromine batteries which can release toxic chlorine (or bromine) and zinc chloride (or zinc bromide) fumes when exposed to fire.

In addition to the environmental problems associated with the malfunction of the advanced batteries now under development, the manufacture of such batteries will require the disposal of waste materials such as lithium, sodium, lead, zinc compounds, etc., which will have a direct impact on local water quality, landfills, and the local ecology unless suitable disposal techniques and procedures are developed and implemented.

Hydrogen Storage

Hydrogen holds promise as a versatile and efficient energy carrier. Because of its flexibility, hydrogen is considered as the fuel of the future by many authorities, and hydrogen energy storage systems are being increasingly recognized and developed for a variety of applications (Dickson *et al.,* 1975; Lotker *et al.,* 1974).

Many techniques are available for the production of hydrogen. Electrolysis of water is the simplest, cleanest, and best known of these techniques (Konopka and Gregory, 1975; Allison and Hughes, 1975). Other processes under development include thermochemical water splitting (using heat from nuclear reactors), steam hydrocarbon reforming, radiolysis of water in aqueous reactors, and the use of photosynthetic organisms and the photolysis of water by sunlight.

Fig. 7.4 shows the basic components of a hydrogen energy storage and reconversion system employing electrolysis cells for hydrogen production. Electrical energy from a variety of sources (off peak utility power, electricity generated by solar cells and wind-electric systems, and electricity generated from ocean thermal gradient systems are some of the commonly suggested inputs) can be used to dissociate water into hydrogen and oxygen. This process can be made very efficient (above 90%) by operating the cells at high pressures (1000 to 3000 psig)* and moderate temperatures (177 to 200°C). As illustrated in Fig. 7.5, the ideal reversible voltage of the hydrogen-oxygen reaction decreases with

* Pound force per square inch $lbf/in^2(g)$ = psig: gauge pressure 1 $lbf/in^2 = 6{\cdot}895 \times 10^3$ Pascal (Pa), the SI unit for pressure.

increasing temperature and increases with increasing pressure. However, in practice, an increase in operating pressure also has the effect of decreasing the operating voltage of the cell. Consequently, performance characteristics of electolysis cells are enhanced when the environmental pressure and temperature associated with such cells are increased (Ramakumar, 1968).

Fig. 7.4 Hydrogen Energy Storage System.

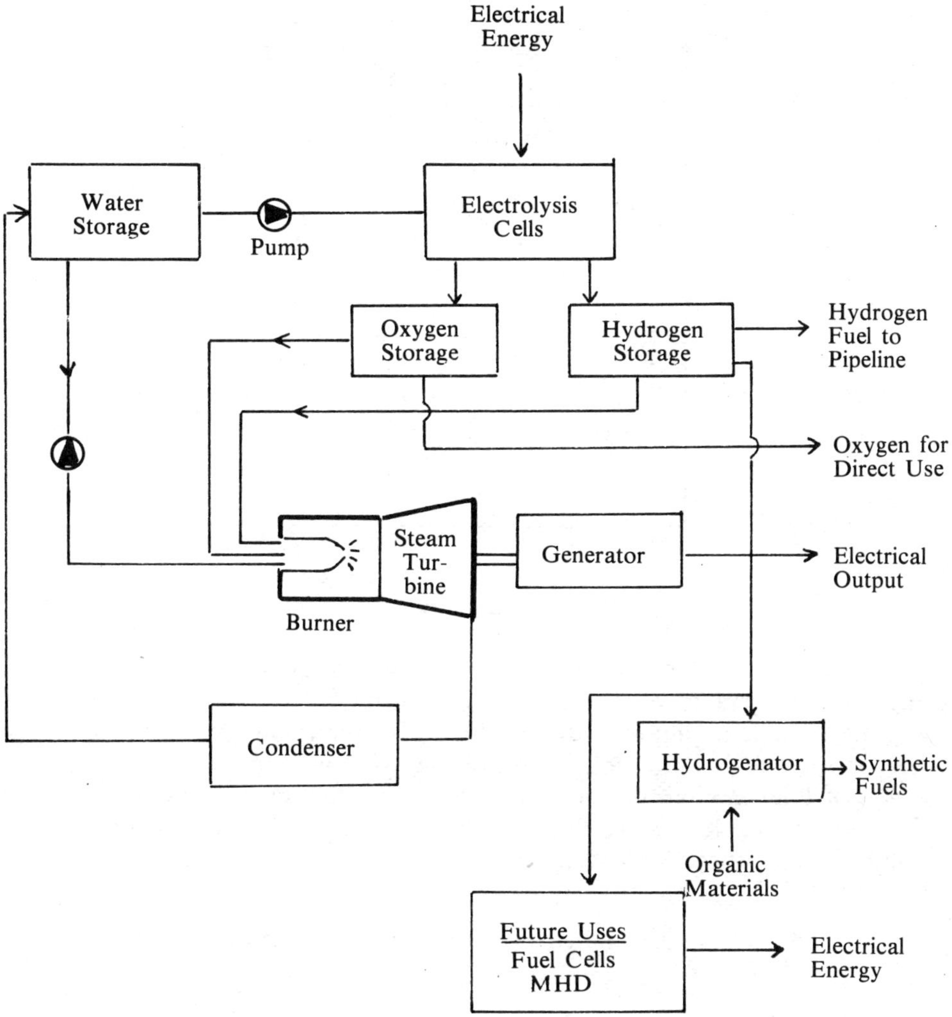

Various alternatives available for the utilization of stored gases (hydrogen and oxygen) are shown in Fig. 7.4. The potential variety of output energy forms makes the concept of a "hydrogen economy" of significant interest to the scientific community.

Fig. 7.5 Ideal Reversible Voltage of Hydrogen Oxygen Reaction Forming Liquid Water (After: Allison, 1976).

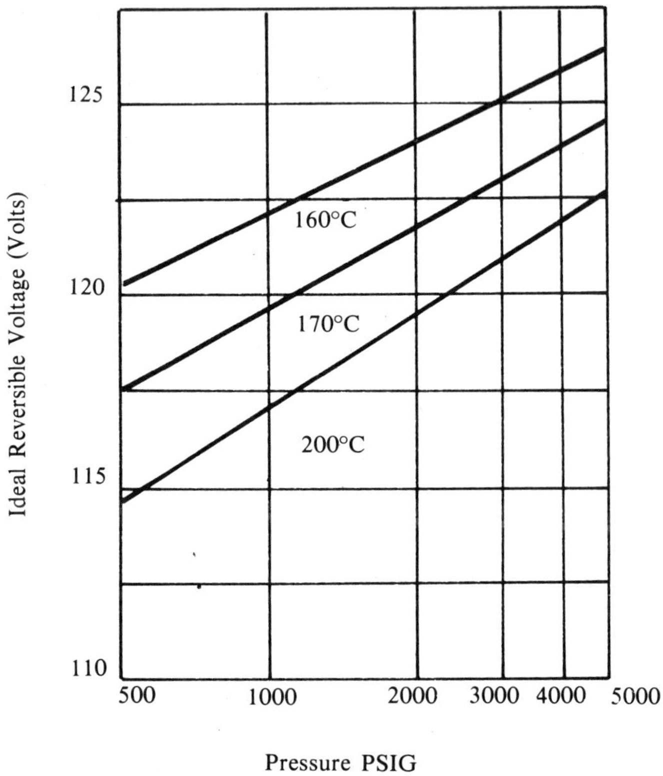

Various options are presently under examination for hydrogen storage. Storage in the form of compressed gas requires large and heavy containers. Cryogenic storage as liquid hydrogen involves continuous refrigeration. Metal hydride storage techniques appear promising (especially for mobile application) in terms of high storage density and long term stability without refrigeration (Strickland *et al.,* 1974). Table 7.2 summarizes the specific energy density characteristics of these hydrogen storage systems.

Table 7.2. Estimated Hydrogen Storage Comparisons

	Wh/kg	kWh/m^3
Hydrogen gas (2,000 psi)	240	371
Liquid Hydrogen	25,872	2,225
Iron Titanium Hydride	530	1,723
Magnesium Nickel Hydride	2,157	1,783

Advocates of hydrogen as a fuel (or energy carrier) for the future stress the fact that burning hydrogen in air does not contribute to the CO_2 buildup in the atmosphere. Water vapor produced during this oxidation reaction can be expected to have some impact on local weather conditions, however, depending on the size of the system. Uncertainty remains as to the environmental impact of burning hydrogen in air in terms of side reactions involving atmospheric nitrogen. These uncertainties could be removed by burning hydrogen in an oxygen environment. However, the environmental benefits of this approach will probably be offset by the need to store or transport oxygen to make this approach possible.

Large scale production of hydrogen can be expected to have environmental consequences. Electrolysis cell failure can result in the discharge of the electrolyte into the environment, and improperly handled discharge from water pretreatment plants can contaminate local water resources. Problems associated with hydrogen embrittlement pose additional hazards when hydrogen is transported by pipeline because of fire or explosive hazards which may occur under conditions of pipeline failure.

Synthetic Fuels

One technique which has been suggested as a means to minimize the problems associated with hydrogen storage involves using the hydrogen to synthesize and store an array of hydrocarbons and alcohols (R & D Goals Study, 1972). Ammonia, hydrazine, methane, methanol, ethanol and gasoline represent some of the synthetic fuels now under consideration and development. Many experts believe that long-term requirements for gaseous and liquid fuels are most likely to be satisfied by synthetic fuels obtained from non-fossil sources.

Producing fuels from agricultural products obtained through intensive agriculture has been proposed as a method of storing solar energy (Alich and Inman, 1975; Kemp and Szego, 1974). Fermentation of grains and starches can yield ethanol and wood, and organic matter can be used directly as fuel or hydrogenated to produce gaseous or liquid fuels. Mounting pressures of global fuel supplies and the need to grow more food with existing arable land tend to make these proposals applicable only under very special circumstances.

Rechargeable Fuel Cells

A rechargeable fuel cell is one in which anode and cathode reactants can be restored by the energy derived from a supplementary energy source (Hughes *et al.,* 1970). This energy source can be thermal, electrical, or some type of radiation source, including solar insolation. With the hydrogen-oxygen reaction, the same set of electrodes can be employed for both electrolytic and fuel cell modes of operation. Turn around efficiency can be as high as 55%. Major technical, economic and reliability problems must be solved before such cells become practical.

THERMAL ENERGY STORAGE

At present there are three basic approaches to storing thermal energy: sensible heat storage, in which a liquid or solid is heated without melting or changing state; latent heat storage, which involves heating a material which melts, vaporizes, or undergoes some other change of state at an essentially constant temperature; and thermo-chemical energy storage, which involves the use of heat to produce a chemical reaction that can be reversed to release thermal energy on demand (Telkes, 1975; Meyer and Hausz, 1975).

Sensible Heat Storage

This is the simplest and most common thermal energy storage system. It is used primarily for storing relatively low temperature thermal energy, and liquid water and/or rocks are normally used as storage materials for such systems. The relatively low temperature associated with this technique (generally below 150°C) generally limits its application to space heating, because conversion to other forms, mechanical or electrical, is low and limited by fundamental laws of thermodynamics.

Small, decentralized sensible heat storage systems represent little threat to the environment. Most small systems use an insulated hot water tank located in an equipment room or buried in the ground. Properly designed systems will be completely disconnected from potable water lines or contain thermosiphon valves which will present backflow from the storage system into the water line. Such precautions are necessary because algae, fungi, and other contaminants can occur in the stored water.

Researchers have investigated the use of contained underground aquifers for large scale sensible heat storage. This process is illustrated in Fig. 7.6. Hydrological characteristics and groundwater flow patterns can be affected by this process, which requires the cycling and storage of heated water in underground reservoirs. Such cycling can also affect the stability of adjacent rock formations and cause leaching of salts from the aquifer. Such leaching could result in a contamination of the aquifer and local groundwater and land resources.

Solar ponds represent another example of sensible heat storage systems. They act as a collector and a storage system for solar energy. The salt gradient pond is the most advanced type of solar pond. It contains a stratified salt solution which increases in salt concentration with increasing pond depth. This stratification minimizes thermal buoyancy convection, resulting in a system the temperature of which increases with increasing depth. Fig. 7.7 illustrates the general character of a stratified salt solar pond. Systems of this type have demonstrated a capability of operating at storage layer temperatures just below the boiling point of water.

Fig. 7.6 Use of Aquifers for Thermal Storage (After: Hanz, 1975).

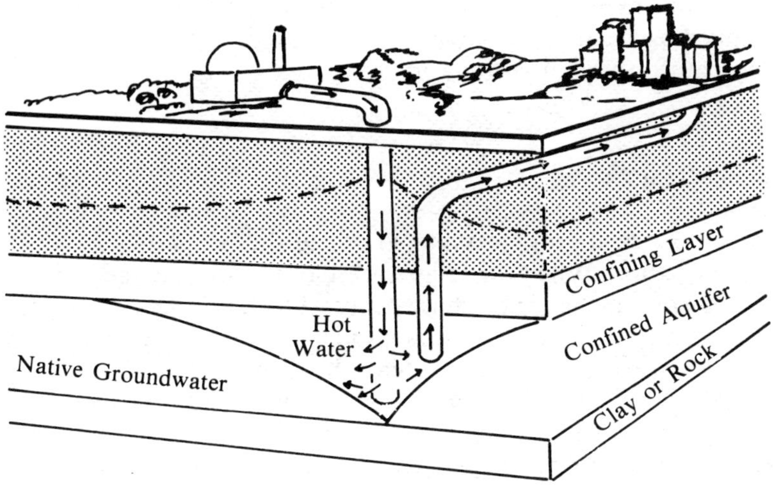

Fig. 7.7 Salt-Gradient Solar Pond (After: Jayadev, 1980).

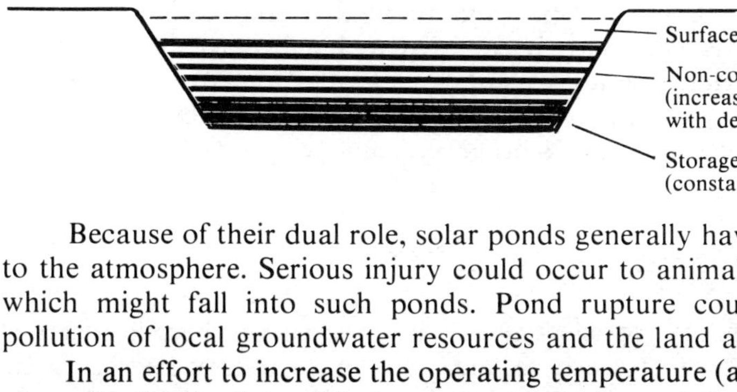

Because of their dual role, solar ponds generally have their surface exposed to the atmosphere. Serious injury could occur to animal or human populations which might fall into such ponds. Pond rupture could result in significant pollution of local groundwater resources and the land area bounding the pond.

In an effort to increase the operating temperature (and, therefore, the range of applications) of sensible heat storage systems, researchers have considered using several other materials in such systems. Organic and inorganic heat transfer oils, molten salts, and liquid metals such as sodium are some materials under consideration. Operating temperatures between 300°C and 800°C appear feasible for such materials (the materials are listed in order of their potential operation at higher temperatures). Some of the heat transfer oils can decompose from thermal cracking when used in sensible heat storage systems. Volatile gases such as methane and ethane can be produced from such a decomposition reaction and cause abrupt increases in system pressure which can result in accidental oil release. Some of the environmental hazards associated with such a release are: skin and eye irritation, burns, toxic vapor inhalation, fires, and physical damage to surrounded area caused by the spread of hot oil.

Molten mixtures of inorganic salts are highly reactive and corrosive. They can decompose at high temperatures, releasing NO, NO_2 and other oxides. Liquid metals such as sodium are even more reactive and corrosive than inorganic salts, thus increasing the environmental risks associated with accidental release of high temperature thermal storage material.

Latent Heat Storage

Latent heat storage systems utilize the properties of an isothermal phase transformation. When the temperature of the storage medium is increased to the melting point a change of phase occurs which absorbs the latent heat of fusion of the material. This heat can subsequently be released when thermal energy is extracted from the storage system, through a reversal of this phase change. A significant advantage of such a process over sensible heat systems in that thermal energy can be withdrawn at an essentially constant temperature. High energy densities and potentialergy densities and potential advantages of latent heat storage systems.

A variety of fluorides, chlorides, phosphates, sulphates, nitrates, nitrites, and hydroxide eutectics are currently under consideration as material for latent heat storage systems. Material characteristics of interest include heat of fusion, heat capacity, thermal conductivity, toxicity and thermal decomposition rate. At present, all of the materials which have been tested are highly corrosive, and most tend to decompose at high temperatures. Other environmental effects associated with latent heat storage systems are similar to those which have been discussed for the more advanced sensible heat storage systems.

Thermochemical Energy Storage

Thermochemical energy storage systems involve an endothermic reaction for energy storage and a reverse exothermic reaction for energy release. Such systems share the advantage of producing heat at an essentially constant temperature with latent heat systems. Thermochemical systems have the additional advantage of not requiring insulated storage tanks. Potentially, thermochemical energy storage systems can allow storage densities several times higher than other types of thermal storage systems.

Partially offsetting the above advantages is the fact that thermochemical storage systems require complex and sophisticated reaction apparatus which must be carefully maintained by trained personnel. Consequently, such storage systems will be applicable only to relatively large systems. Some of the chemicals considered for use in thermochemical energy storage systems are listed in Table 7.3. Potential environmental problems associated with their use are suggested in the Table.

Table 7.3. Description of Chemicals Considered for Use in Thermo-chemical Energy Storage Systems (after Casarett and Doull, 1975)

Chemical	Characteristics/Description/Uses
$Mg(OH)_2$	Magnesium hydroxide — classified as a nonsystemic antiacid, main component of "Milk of Magnesia" suspension, low solubility
MgO	Magnesium oxide — hydrates readily to magnesium hydro-oxide and has similar uses, O.E.* Fumes is 15,000 ug/M^3
CaO	Calcium oxide — hydrates readily to calcium hydroxide component of cement and wall plasters, may cause alkali burns to eyes
$Ca(OH)_2$	Calcium hydroxide — sparingly soluble, cause severe lime alkali burns in the eyes, very difficult to remove once in the eyes
CH_4	Methane — combustible, physiologically inert, simple asphyxiant (prevent body tissues from receiving adequate oxygen when inhaled at 20 — 30% by volume)
CO	Carbon Monixide — Chemical asphyxiant — has a 200 times greater affinity for hemoglobin, the blood's oxygen carrier, 0·1%, 1-hour exposure dangerous to life, 0·35% exposure fatal in less than one hour
CO_2	Carbon Dioxide — minor atmospheric component, respiratory gas, simple asphyxiant
H_2SO_4	Sulphuric acid — severe eye damage at pH greater than one, chronic and acute effects on pulmonary function for sulphur particulates, conc. H_2SO_4 is a strong acid and destroys tissue on contact
SO_3	Sulphur trioxide — reacts rapidly with moisture to form sulphuric acid
SO_2	Sulphur dioxide — causes eye irritation (10 ppm), lung irritation and severe coughing (20 ppm), affects pulmonary function

* O.E. signifies Occupational Exposure limits for 8-hour exposure.

ELECTRICAL STORAGE

At the present state of technology, energy storage in electric fields is not practical except for storing small amounts of energy as fractions of an AC cycle using capacitors. For such applications, present day capacitors are economically viable and are used extensively by utilities for power factor correction.

Magnetic Fields

Progress in superconductive technology, coupled with the emergence of reliable solid-state converter (rectifier/inverter) systems has made it possible to store large amounts of energy in high-intensity magnetic fields set up by superconductive inductors (Peterson *et al.,* 1975; Hassenzahl *et al.,* 1973). A simplified schematic of this configuration is shown in Fig. 7.8. Interface with a three-phase power system is easily accomplished by appropriate gating control of the solid state thyristors in the converter, both during the charge and discharge modes. During operation, the current through the inductor can vary over a 2 to 1 range corresponding to an energy variation of 4 to 1 or a 75% energy extraction.

Fig. 7.8 Superconducting Energy Storage System.

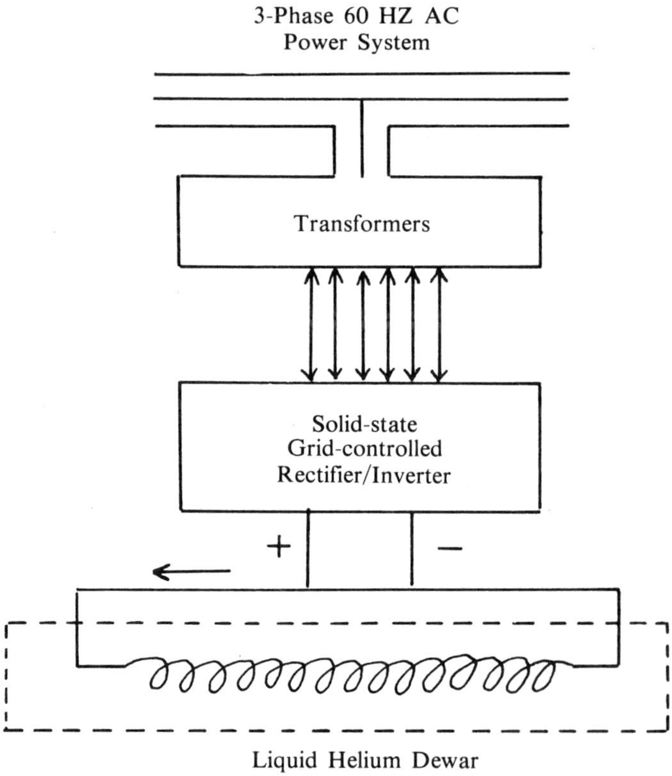

Studies have shown that superconducting inductors can effectively dampen electro-mechanical oscillations in power system and improve their stability. The fact that they represent a zero resistance conductor has made them attractive to electric distribution planners. Superconducting systems can be located underground, and they are potentially the most efficient energy storage now under consideration (efficiency can approach 100%).

Key environmental concerns for superconducting technology are associated with the consequences of possible seismic events or major accidents which could result in the release of cryogenic fluid. The release of large quantities of supercool fluid represents a serious threat to those who work near such systems, and the spontaneous release of energy stored in the magnetic field of the superconducting system can be potentially destructive over a large area — especially in large systems.

CONCLUDING REMARKS

None of the energy storage techniques which have been discussed has reached the stage of development required for wide spread application. All have environmental problems which must be addressed before they can be expected to contribute significantly to the power systems of the future. Concentrated energy in any form is potentially hazardous to the local environment. The hazards associated with energy storage are directly related to system size. Small, decentralized energy storage units can be expected to be relatively benign in terms of environmental problems and safety hazards, when properly designed and maintained. Large, centralized energy storage systems of the type envisioned for load smoothing in electrical power systems will require extensive research and development before their environmental characteristics can be properly documented.

Energy storage by pumped hydro is a mature technology and its economics are well established. Underground pumped hydro costs are strongly dependent on the existence of natural caverns and man-made mine shafts. Compressed air-storage and gas-turbine plants appear to be economically viable at present. Isothermal hydraulic compression has not been used for nearly three-quarters of a century. The inherent simplicity and possible high efficiency of isothermal hydraulic air compression suggests that this technique should be re-examined in light of modern technology. The requirement of a suitable terrain would limit its applicability to selected hilly regions with an abundant supply of water.

Projected costs of superflywheel systems appear attractive for short-term (2 to 3 hours) energy storage, to provide spinning reserve capacity in power systems. They may also be adaptable for use with wind energy systems to smooth out and exploit the energy in wind gusts.

Massive research efforts underway in the development of advanced secondary batteries promise a variety of economically viable battery systems for large-scale and small-scale energy storage within the present decade. Required AC/DC interfaces will not pose any serious problems for the adoption of sub-station size systems with storage capacities in the hundreds of MWh. The use of such batteries in electric vehicles is also anticipated.

In the foreseeable future, thermal storage is expected to be only on a small scale, in individual homes and buildings, to harness solar energy. The economics of such systems are presently attractive in regions where conventional fuel costs are high.

Superconducting magnetic energy storage systems are economically feasible only on a large scale (1,000 MWh and above). They can be deep discharged without damage and as such are ideally suited for short-term (2 to 3 hours) storage for levelling daily peak load variations.

The economics associated with hydrogen and synthetic fuels production appear to be promising, but precise data concerning the future costs of such fuels (storage systems) is presently unavailable.

In summary, energy storage is a very active and changing field. With the number of talented people and active research programs involved, it is only a matter of time before economical energy storage systems in a multitude of technologies and sizes are available for use in alleviating the burdens presently imposed on the energy resources of the world.

References

Alich, J. A. Jr. and R. E. Inman (1975): Energy from agriculture. Record of the Tenth IECEC, Newark, Delaware, pp. 834–841.

Allison, H. J. and W. L. Hughes (1975): Energy storage by high-pressure moderate-temperature electrolytic techniques. Record of the Tenth IECEC, Newark, Delaware, pp. 104–110.

Allison, H. J. (1976): A wind energy system utilizing high pressure electrolysis as a storage mechanism. Proceedings of the First World Hydrogen Energy Conference, Miami Beach, Florida, pp. 2B-3 to 2B-14, March 1976.

Casarett, L. J. and J. Doull (1975): Toxicology. Macmillan, New York.

Day, W. H. *et al.,* (1974): Pumped air storage for electric power generation. IEEE Power Engineering Society Paper, Energy Development, Catalog No. 74 CHO 913-4-PWR, pp. 30–36.

Dickson, E. M. *et al.,* (1974): The use of personal vehicles within the hydrogen energy economy — an assessment. Proceedings of the Hydrogen Economy Miami Energy (THEME) Conference, Miami, Florida, pp. S8-1 to S8-14, March 1974.

Dickson, E. M. *et al.,* (1975): A technology assessment of the hydrogen economy concept. Record of the Tenth IECEC, Newark, Delaware, pp. 1370–1379.

Douglas, D. L. (1974): The impact of advanced batteries on electric power generation. IEEE Power Engineering Society Paper, Energy Development, IEEE Catalog No. 74 CHO 913-4-PWR, pp. 37–41.

Fernandes, R. A. (1975): Optimum peak-shaving mix for electric utilities. IEEE Conference Paper No. C 75 143-3, presented at the Winter Power Meeting, New York, NY.

Fullman, R. L. (1975): Energy storage by flywheels. Record of the Tenth Intersociety Energy Conversion Engineering Conference (IECEC), IEEE Catalog No. 75 CHO 983-7 TAB, Newark, Delaware, pp. 91–100.

Hanz, W. (1975): Energy Conservation: is the heat storage well the key? Public Utilities Fortnightly, April 24, p. 34.

Hassenzahl, W. V. *et al.,* (1973): The economics of superconducting magnetic energy storage systems for load levelling. Los Alamos Scientific Laboratory Informal Report No. LA-5377-MS, August 1973.

Hughes, W. L. *et al.,* (1962): Energy storage-key to our economic future. Oklahoma State University, Office of Engineering Research, Stillwater, Oklahoma.

Hughes, W. L. *et al.*, (1970): An economical energy storage system utilizing hydrogen-oxygen regenerative fuel cells. IEEE Conference Paper No. 70 CP 212-PWR, presented at the IEEE Winter Power Meeting, New York, NY, January 1970.

Jayadev, T. S. (1980): Solar ponds and their applications. Proc. 7th Energy Technology Conference, Washington D.C. p. 1,194.

Kemp, C. C. and G. C. Szego (1974): The energy plantation. Paper presented at the IEEE 1974 International Conference on Systems, Man and Cybernetics, Dallas, Texas, September 1974.

Konopka, A. J. and D. P. Gregory (1975): Hydrogen production by electrolysis: present and future. Record of the Tenth IECEC, Newark, Delaware, pp. 1,184–1,193.

Lawson, L. J. (1975): The kinetic energy wheel: a means of reducing pollution while conserving resources. Proceedings of the 21st Annual Meeting of the Institute of Environmental Sciences, Vol. 1, Anaheim, California, pp. 130-133, April 1975.

Lotker, M. *et al.*, (1974): The hydrogen economy — a utility perspective. IEEE Conference Paper No. C 74 099-8, Energy Development, IEEE Catalog No. 74 CHO 913-4-PWR, pp. 89–97.

Meyer, C. F. and W. Hausz (1975): A new concept in electric generation and energy storage. Proceedings of the Frontiers of Power Technology Conference, Stillwater, Oklahoma, pp. 17-1 to 17-17, October 1975.

Peterson, H. A. *et al.*, (1975): Superconductive energy storage inductor-converter units for power systems. IEEE Transactions on Power Apparatus and Systems, Vol. PAS-94, No. 4, pp. 1,337-1,348, July/August, 1975.

R. & D. Goals Study (1972): Hydrogen and other synthetic fuels. A summary of the work of the synthetic fuels panel, prepared for the Federal Council on Science and Technology R & D Goals Study, U.S. Government Printing Office, No. TID-26, 136, Washington, D.C.

Rabenhorst, D. W. (1973): Superflywheel energy storage system. Proceedings of the Workshop on Wind Energy Conversion Systems, sponsored by NSF/NASA, Document Number NSF/RA/W-73-006, Washington, D.C., pp. 137–145.

Ramakumar, R. (1968): An experimental hydrogen-oxygen regenerative fuel cell. Proceedings of the Frontiers of Power Technology Conference, Oklahoma State University, Stillwater, Oklahoma, pp. 18-1 to 18-36, October 1968.

Ramakumar, R. *et al.*, (1969): A wind energy storage and conversion system for use in underdeveloped countries. Proceedings of the 4th IECEC, Washington, D.C., pp. 606–613.

Ramakumar, R. *et al.*, (1974): Prospects for tapping solar energy on a large scale. Solar Energy Journal, Vol. 16, No. 2, pp. 107–115.

Ramakumar, R. (1975): Continuous duty solar energy system concepts. Record of the Tenth Intersociety Energy Conversion Engineering Conference (IECEC), IEEE Catalog No. 75 CHO 983-7 TAB, Newark, Delaware, pp. 759–764.

Ramakumar, R. *et al.,* (1975): Solar energy conversion and storage systems for the future. IEEE Transactions on Power Apparatus and Systems, Vol. PAS-94, No. 6, pp. 1926–1934.

Ramakumar, R. (1976): Utilization of solar and wind energy to improve the living environment in less developed countries. Proceedings of the 22nd Annual Meeting of the Institute of Environmental Sciences, Philadelphia, PA, April 1976.

Rogers, F. C. and A. E. Allen (1974): Energy storage underground. IEEE Power Engineering Society Paper, Energy Development, Catalog No. 74 CHO 913-4-PWR, pp. 21–29.

Salihi, J. T. (1973): Energy requirements for electric cars and their impact on electric power generation and distribution systems. IEEE Transactions on Industry Applications, Vol. IA-9, No. 5, pp. 516–532.

Scott, F. M. (1977): Underground Hydro-electric pumped storage: a practical option. Energy, p. 20.

Strickland, G. *et al.,* (1974): An engineering-scale energy storage reservoir of iron titanium hydride. Proceedings of the THEME Conference, Miami, Florida, pp. S4-10 to S4-21, March 1974.

Telkes, M. (1975): Thermal energy storage. Record of the Tenth IECEC, Newark, Delaware, pp. 111–115.

Woodbridge, D. E. (1970): The hydraulic compressed-air power plant at the Victoria mine. The Engineering and Mining Journal, Vol. LXXXIII, No. 3, pp. 125–130.

Yao, N.P. and J. R. Birk (1975): Battery energy storage for utility load levelling and electric vehicles: a review of advanced secondary batteries. Record of the Tenth IECEC, Newark, Delaware, pp. 1107–1119.

CHAPTER 8

Energy from Biomass

ESSAM EL-HINNAWI
United Nations Environment Programme, Nairobi, Kenya
and
FATMA EL-GOHARY
National Research Centre, Dokki, Cairo, Egypt

Fig. 8 "Cow-dung Cakes"

THE RESOURCE BASE

Biomass is defined as renewable organic matter produced by photosynthesis – a mechanism that converts solar energy into stored chemical energy – directly in the case of plants and indirectly in animals whose ultimate food source is plant material. Biomass can be trees, crops, aquatic plants, and organic wastes of different kinds. Biomass also includes fuel crops specifically cultivated on energy farms for their fuel content (see section on energy farms). Fuel crops could be conventional annual crops, fast-growing trees, aquatic plants, or even multi-plant communities. Biomass is an important source of energy; for it is the principal fuel (e.g. fuelwood and cow-dung cakes) for the vast majority of people in most developing countries.

Photosynthesis is the key process in life and as developed by plants can be simply represented as:

$$H_2 + CO_2 \xrightarrow[\text{solar energy}]{\text{plants}} \text{Organic material} + O_2$$

In addition to C, H and O, plants also incorporate nitrogen and sulphur into organic material via light-dependent reactions. Photosynthesis is theoretically and practically the least efficient of any of the direct solar energy conversion processes. Photosynthetic conversion efficiencies vary from less than 0.5 to about 5% and represent the efficiency of the total process, sunlight-fixed chemical energy (Hall, 1979; Benemann, 1978). Under optimum field conditions values between 3 and 5% conversion are achieved by plants; however, often these values are for short-term growth periods, and when averaged over the whole year they fall between 1 and 3%. In practice, photosynthetic conversion efficiencies in temperate areas are typically between 0.5 and 1.3% of the total radiation when averaged over the whole year, while values for sub-tropical crops are between 0.5 and 2.5% (Hall, 1979); the average for the biosphere is about 0.25%. Hall (1979) estimates that each year plant photosynthesis fixes about 2×10^{11} tonnes of carbon with an energy content of 3×10^{21} J; this is about 10 times the world's annual energy use and 200 times our direct food-energy consumption, even though the photosynthetic process is operating at only a 0.1% efficiency. Sørensen (1980) estimates that the average conversion of solar energy by the biosphere is 133×10^{12} W of which 76×10^{12} W is terrestrial, and 57×10^{12} W is marine. The dry weight of all living biomass (plant matter) on the earth's land surface has been estimated as nearly $2{,}400 \times 10^9$ dry tonnes, with a primary production rate of 172×10^9 dry tonnes/y (Rodin *et al.*, 1975). The latter is equivalent to more than 10 times the present total world energy consumption from all commercial sources (assuming that the average calorific value of dry biomass is about 1900 kJ/kg). However, only a modest fraction of the biomass enters the human economy as waste or residues.

Trees are an important source of biomass. Forests and woodland cover approximately 30% of the world land area – 3.800 million ha., of which 1,460 million ha. are classified as tropical forests, 220 million ha. as sub-tropical forests, 1,000 million ha. as open savannah woodlands, 450 million ha. as temperate forests and the remaining 670 million ha. as boreal forest. Another source of biomass is grass. Earl (1975) estimated the world's grassland area to be 26×10^8 ha. In the tropics, as in other regions, a sizeable fraction of the natural grass biomass is grazed by wild animals and by domestic livestock. Another and smaller fraction is harvested for use as fodder. Highly productive grass, for example, elephant grass (*penonisetum purpureum*) offers great possibilities as a source of forage for domestic animals and for energy (in case of energy farms discussed later). However, the extensive development of this source of biomass is still being studied to determine its viability as a source of energy.

Practically all agricultural crops produced by farmers throughout the world are technically capable of being converted into some form of energy, whether the primary purpose of the crop be for food, animal feed, fibre, or other products. However, the emphasis for energy use at present has focussed on some specific crops, for example, sugar cane, cassava, etc. and on agricultural and agro-industrial residues. Crop residues vary and range from none in alfalfa to 6.7 tonne per hectare in corn (air dry weight). Only corn and rice exceed the crop average of 4.4 tonne of residue per hectare (Pimentel *et al.,* 1978). Throughout the world, farm crops leave substantial residues, the extent and scale of which is rarely realized. Wheat, with a yearly crop production of 355 million tonnes, sorghum with 55 million tonnes, millet with 46 million tonnes, and several other less widely-grown grain crops all contribute to a grand annual total of 1,700 million tonnes of cereal straw (UNEP, 1978), much of which is at present unutilized. Agro-industries also produce vast quantities of residues. The sugar cane industry, for example, creates each year 50 million tonnes of sugar cane tops and 67 million tonnes of bagasse, as well as molasses and press mud. Although wastewater from molasses is used in Indonesia to fertilize rice paddies, the press mud is widely used as a soil conditioner and as a component of animal feed, and the bagasse is often burned as a low-grade fuel in the sugar mills themselves, the sugar cane residues are generally greatly under-utilized. Pineapple is typical of many fruits for which much of the crop is wasted. Where pineapples are canned, less than 20% of the whole fruit is used; the remainder, often in the form of highly-polluting liquid, can cause considerable disposal problems. In Malaysia in 1974, for example, 250,000 tonnes of pineapples generated only 40,000 tonnes of canned fruit, but 210,000 tonnes of residues. The core, skins and fresh wastes of pineapples are often crushed for juice, and in Kenya and elsewhere the resulting bran is dried and used as cattle feed. In the Philippines, the residues are converted into wine but in other countries they are not utilized at all (UNEP, 1978).

Agricultural residues are used extensively in many rural areas of the developing countries as a source of energy (through direct combustion). It is indeed difficult to give an estimate of the amount of agricultural residues used as fuel, since they are of non-commercial nature and have not so far been accounted for in any global, regional or national statistics. Earl (1975) estimated that the world annual consumption of agricultural residues, e.g. bagasse, cotton sticks, and hay, as fuel is about 10 million tonnes of coal equivalent. However, Zumer-Linder (1976), referring to India alone, quotes an estimate of the consumption of agricultural residues as totalling 27 million tonnes/y (or about 17 million tonnes coal equivalent). A wide range of proposals exists concering the utilization of crop remains as a source of energy, but analysis of the agricultural, environmental, and energetic aspects of the use of crop remains suggest that little or none of these remains should be used for biomass energy conversion (Pimentel *et al.,* 1978; Chartier, 1979). Crop remains left on land prevent sediment runoff, conserve soil and water, maintain soil carbon ratios and soil structure, and prevent nutrient (N, P, K, etc.) loss. The beneficial role of crop residues in agriculture, including their potential as feed for livestock, outweigh their use as potential fuel. On the other hand, food processing wastes (pods, hulls, cobs, chaff, and other by-products) have value as an energy source (UNEP, 1978).

Aquatic vegetation (fresh water marine) has recently attracted attention as a source of biomass for energy conversion. In the warm water of the tropics aquatic weeds exist and multiply in abundance and have been used to a varying extent for food, animal feed and fertilizer. However, in many countries, the nuisance value of water weeds exceeds their benefits: they impede navigation, irrigation and fish culture projects and may contribute to the spread of tropical diseases. Because most acquatic plants contain large quantities of water (about 90%), they are unsuitable for most energy-conversion processes unless they are predried by pressure (which itself is energy-consuming) or by spreading in the sun (which is time and space-consuming). At present, anaerobic fermentation appears to be the only feasible process for converting this biomass into fuel.

The only biomas of animal origin that has any significance for energy production is excrement. Besides its great importance as a fertilizer, animal dung (dung-cakes) has been used for centuries as fuel in many developing countries. Earl (1975) suggests that the total world consumption of cow dung for fuel is not less than 150 million dry tonnes/y of which 40% is in India; other estimates vary from 80 million tonnes/y to 200 million tonnes/y. As in the case of agricultural residues, it is difficult to estimate the amounts of dung used as fuel, since most of it is non-commercial, and has not been accounted for in any national statistics.

Urban refuse or municipal solid wastes constitute also an important source of biomass. The characteristics and quantity of the urban refuse are factors in not only the amount of energy that can be recovered, but also in the type of energy recovery system that can be employed. The quantity of solid waste generated in most industrial countries is substantial. In the U.S.A., for example, about 115×10^6 tonnes of urban refuse are generated annually; in the UK, about 15×10^6 tonnes/y while in Spain the figure is about 9×10^6 tonnes/y (Pfeffer, 1979). Assuming that the energy content of U.S. refuse is about 11 GJ per tonne, the annual amounts of refuse in the U.S.A. could produce 1.3×10^{18} J/y.

Biomass as an energy source has positive and negative effects on the environment. On the positive side, it acts as a renewable energy source, as a sink for atmospheric carbon dioxide and it contributes to soil stabilization, reduction of water run off and desertification (El-Hinnawi, 1981). Biomass provides a low-sulphur fuel and an inexpensive source of energy, especially for rural areas of developing countries. The conversion of organic wastes into fuel, in addition to providing a source of energy, reduces the environmental hazards associated with these wastes. On the other hand, if biomass is not properly managed, soil erosion, desertification and related detrimental environmental consequences will occur. Moreover, the land used for biomass production may be partially lost for other uses – such as for food production.

A number of technologies, at different stages of development, exist for the conversion of biomass into fuel. These are summarized in Fig. 8.1. The most commonly used at present are direct combustion, pyrolysis, anaerobic digestion and fermentation.

Fig. 8.1 Conversion of Biomass

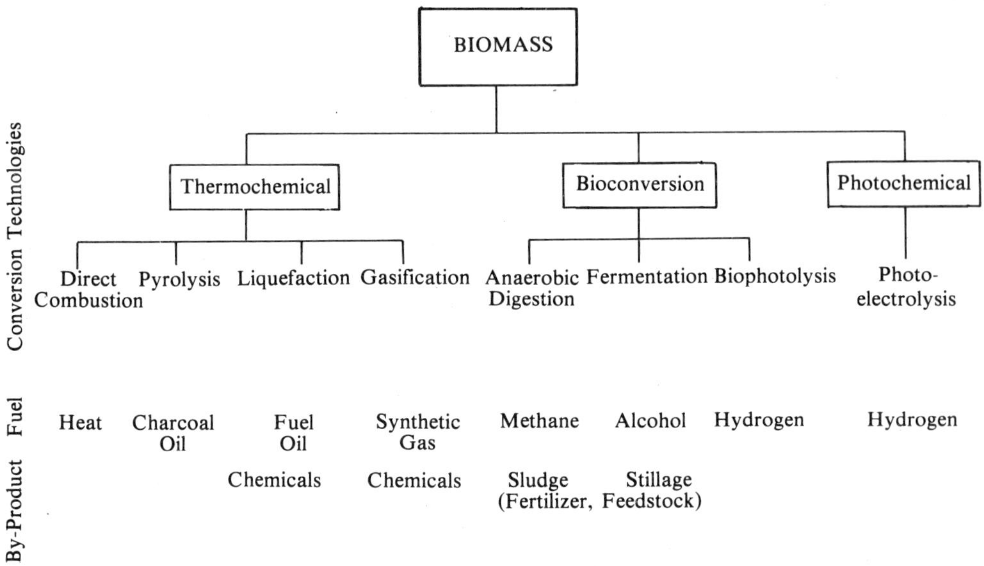

FUELWOOD AND CHARCOAL

Wood has been a primary fuel since man discovered fire. Wood is composed principally of cellulose and ligno-cellulose together with gums, resins, inorganic matter, and a variable amount of moisture, the amount depending on the kind of wood, the season in which it is cut, and the extent to which it has been allowed to dry. The approximate proportions by weight of the main elements present in wood are: carbon 50%, hydrogen 6%, oxygen 44%, and a trace of nitrogen.

Fuelwood can be obtained from any tree whether occurring naturally or planted, either directly from the forest or from waste material produced at sawmills and wood-using industries. The most important controllable factor influencing the efficiency of wood as a fuel is the moisture content; the less the moisture content, the higher the calorific value would be. The gross calorific value of oven-dry wood averages 19.7 kJ/g, but in practice it ranges from 14.7 to 16.8 for wood dried naturally over a period of 4–8 months.

The magnitude of fuelwood consumption in the world, especially in the developing countries, as well as its significance in the everyday life of well over two billion people, was not realized until recently. Many developing countries depend upon wood as their major source of fuel. For example, in Kenya where 90% of the population live in rural areas, they depend almost exclusively on fuelwood as their primary source of energy (80% is used for cooking and the balance mainly for heating). Even in urban areas, about 80% of the population depend on charcoal for energy. In Zambia, fuelwood and charcoal constitute about 88% of sources of energy used in rural and urban households. Fuelwood is also the principle source of energy for cooking and other domestic uses in Sri Lanka, Nepal, Thailand and several other countries, and it is an important fuel in the rural areas of all African and Asian countries. In these areas, where wood is readily available, nearly 95% of households use it as a primary source of energy; the per capita consumption varies from 1.3 m^3 to 2.3 m^3/y, with an average of 1.5 m^3/y^2 (Openshaw, 1978). About 50% of the wood consumed as fuel is used for cooking; 30% for domestic heating and the remaining 20% for processing agricultural products, industry, etc.

Fuelwood Supply and Demand

As mentioned earlier, forests and woodlands cover approximately 30% of the world land area. The total volume (stock) contained in this forest area is between 340 and 360 $\times$ 10^9 m^3, which is equivalent to about 270 $\times$ 10^9 tonnes air-dry weight (12% moisture content) or 175 $\times$ 10^9 tonnes of coal equivalent; about 20 times the world's current energy consumption from all sources (El-Hinnawi, 1981). However, this represents the *resource base* or capital and only the annual increment "interest" from this capital can be used without destroying the resource base. Such annual increment has been estimated to be 6,610 million m^3 (Openshaw, 1978, 1980). Official statistics (FAO, 1980) indicate that the total world consumption of roundwood in 1978 was about 2,600 million m^3 (about 47% of which was used as fuelwood and charcoal*. In other words, the annual

* It should be noted that these figures may be considerably underestimated. No accurate records are kept by many countries of self-collected or self-produced products such as fuelwood, charcoal (the so-called non-commercial sources of energy) and these are not accounted for in official statistics. For example, Openshaw (1978) estimated the total world consumption of wood products to be nearly twice the official consumption figures, the largest difference being accounted for by woodfuel which is estimated to be three times greater than reported consumption. Accordingly, Openshaw (1978) estimated the consumption of roundwood in 1976 to be 4,808 million m^3 instead of the official figure of 2,560 million m^3 (FAO, 1980).

increment of the forest resource base would meet the world demand for wood. However, on a world-wide basis all the increment is not being removed, and much of this unremoved increment is in the inaccessible northern coniferous forests of Alaska, Canada and the U.S.S.R. This has led to an over-exploitation of the forests in certain regions to meet the increasing demand for wood, destroying thereby the resource base. Deforestation of tropical forests has been estimated to occur at a rate of 10–12 million ha/y (Saouma, 1978). Most of this deforestation occurs (and will continue to occur) in the developing countries, whose humid tropical forests and open woodlands are steadily being felled and converted to farmland and pasture. This trend is impelled by several forces: the expansion of agricultural frontiers into forested areas in order to supply food as populations increase; the demand for fuelwood and charcoal; the demand for tropical forest products by industrialized nations; and the demand within the developing countries for paper and other forest-derived products as incomes rise (CEQ, 1980).

Table 8.1 gives the consumption of roundwood and of fuelwood in different regions as presented in official statistics. As mentioned before both figures are under-estimated since they do not take into account "non-commercial" production and consumption figures. Openshaw (1978) estimated that the fuelwood consumption in the world in 1976 was about 3029 million m^3 and that by the year 2000 it will reach about 3882–5028 million m^3.

Table 8.1 Consumption of Roundwood and Fuelwood (including charcoal) in 1978 (After FAO, 1980)

	Million m^3		
Region	Roundwood*	Fuelwood**	% of Fuelwood
World	2600	1218	47%
Africa	390	344	88%
N. America	495	18	36%
L. America	272	198	73%
Europe	305	39	13%
Asia	746	534	72%
Oceania	31	7	23%
USSR	361	78	22%

* of the world total about 1135 million m are coniferous and 1465 million m^3 are non-coniferous.
** of the world total, about 170 million m^3 are coniferous and about 1048 million m^3 are non-coniferous.

CONIFEROUS: All woods derived from trees classified botanically as: *Gymnospermae* – e.g. fir (*Abies*), parana pine (*Araucaria*), deodar (*Cedrus*), ginkgo (*Ginkgo*), larch (*Larix*), spruce (*Picea*), pine, chir, kail (*Pinus*), etc. These are generally referred to as softwoods.

NON-CONIFEROUS: All woods derived from trees classified botanically as *Angiospermae* – e.g. maple (*Acer*), alder (*Alnus*), ebony (*Diospyros*), beech (*Fagus*), lignum vitae (*Guiaicum*), poplar (*Populus*), oak (*Quercus*), sal (*Shorea*), teak (*Tectona*), casuarina (*Casuarina*), etc. These are generally referred to as broad-leaved or hardwoods.

With the present rate of deforestation in the developing regions, where fuelwood is most needed (the demand will increase by the year 2000), situations have evolved where fuelwood has become quite scarce. In a recent study by the Secretariat of the United Nations Conference on New and Renewable Sources of Energy (1981), it is stated that, in 1980, situations of acute scarcity involved approximately 90 million rural people in developing countries in areas where even over-exploitation of existing woody vegetation together with agricultural residue, where available, was insufficient to supply fuelwood requirements. Such situations currently prevail in Africa, mainly in the arid and semi-arid areas south of the Sahara, in the east and south-east and in mountainous areas; in Asia, in the Himalayas and the hills of South Asia; and in Latin America, mostly in the Andean Plateau and the arid areas of the Pacific Coast in South America. In addition some 150 million people live in major semi-urban centres situated in rural areas, already in fuelwood deficit. In 1980, deficit situations involved 833 million rural people and 166 million urban dwellers in areas where the population is still able to meet its minimum energy needs, but only by harvesting in excess of sustainable fuelwood supply. The resource is, therefore, inadequate to meet present needs. Those in such situations in Africa amount to 146 million people, mainly in savannah areas in the west, central and south-eastern areas. In North Africa and the Middle East, 70 million rural people live in situations of fuelwood deficit. In Asia, 550 million people in rural areas and small urban centres are affected, mainly in the Indo-Gangetic plains of the Indian sub-continent and in South-East Asia. In Latin America, 82 million rural people live in such deficit situations, mostly in the semi-arid and arid areas. Under current trends those situations will deteriorate to acute scarcity, which will affect the increased populations of the future around the year 2000.

ENVIRONMENTAL IMPACTS OF PRODUCTION AND USE OF FUELWOOD:

Fuelwood Production

Fuelwood comes overwhelmingly from local sources, and this puts growing pressure on the trees, bushes and shrubs near the centres of population. Long before the demand for fuelwood leads to complete destruction of the tree cover, it can have a markedly degrading environmental effect. Excessive pruning of the branches may reduce a tree's capacity for growth; removal of the more easily-felled younger trees may reduce the regenerative ability of the forest; excessive opening of the canopy through the removal of too many trees can render the forest susceptible to damage from wind and sun and can affect wildlife; the removal of all residues, even to the point in some areas of sweeping up the leaves, removes the nutrients that should return to the soil to maintain its fertility; removal of stumps, bushes and shrubs can destroy much of what remains of the soil's protective cover and binding structure. And eventually, the whole forest may be felled and dissappear (UNEP, 1977).

The uncontrolled and indiscriminate collection of fuelwood for cooking and heating, therefore, can have the most serious implications. Until the 1940s, forests had completely disappeared from most of China, because the trees had been cut down for fuel; in recent years, though, this trend has been reversed and vast areas have been successfully re-afforested. In much of India, however, the forests are still rapidly declining (Eckholm 1975). With the growth of human populations, the forests are being cut down faster than they can grow, partly to make room for new farmlands, and partly for use as fuel. As a consequence, the upland area – the watersheds for the great rivers which flow through the plains – are subjected to destructive erosion, while the resulting sediments cause rapid filling of reservoirs and destructive floods downstream severely reduce the cover of cultivable soil and the food which can be produced from it. In the Chittagong Hill tract in south-eastern Bangladesh, about 40% (870,000 hectares) of the total forested area of the country is being rapidly destroyed by illicit timbering and by shifting cultivators who have reduced the traditional 10/year slash-and-burn rotation to 3 years. In the hills of Nepal, the 3.4 million hectares of forested land are being depleted for fuel and cattle feed at an annual rate of more than 3%. This could lead to total depletion within 30 years. Between 1963 and 1971, in Thailand, the total land area covered by healthy forests was apparently reduced from 53% to 39%. The forested area is now estimated as less than 30% – probably about 16 million hectares. In Sri Lanka, the forest cover has been reduced from 44% in 1956 to about 20% at present of the total land area by a combination of clearing for settled agriculture, slash-and-burn cultivation with progressively shorter rotation times, and illegal exploitation of trees for construction, household and industrial fuels (Revelle, 1980).

In semi-arid regions the ecological consequences of fuelwood consumption contribute to the process of desertification. Fuelwood is a scarce and expensive item throughout sub-Saharan Africa, all the way from Senegal to Ethiopia. A manual labourer in Niamey must now spend nearly one quarter of his income on fuel. But the price is higher than even he realizes. The caravans that bring this precious resource into the towns are contributing to the creation of desert-like conditions in a wide band along the desert's edge. Virtually all the trees within 70 km of Ouagadougou have been consumed as fuel by the city's inhabitants, and the circle of land stripped bare for fuelwood is continually expanding (Eckholm, 1975).

The forest clearing and the burning of fuelwood add to the atmospheric carbon dioxide budget. Recently, it has become evident that additional input of non-fossil excess carbon dioxide into atmosphere, due to forest cutting, forest burning, soil management practices, etc., is of the same order of magnitude as the input of CO_2 from fossil fuel (estimated at 5×10^{15} g C/y^2). The burning of fuelwood may inject CO_2 into the atmosphere at a rate of $0.2 - 0.4 \times 10^{15}$ g C/y, and forest clearing in the developing countries is estimated to inject 3.6×10^{15} g C/y (SCOPE, 1979). About half of these quantities of carbon dioxide are removed from the atmosphere through different geochemical cycles, and the net contribution of non-fossil fuel sources to the carbon dioxide budget of the atmosphere is not accurately known but considered a substantial one (Hinckley, 1980 estimated the net contribution of CO_2 to the atmosphere and hydrosphere from deforestation to be about 1.5×10^{15} g C/y^2). The increase in atmospheric CO_2 will lead to an increase in the earth's surface temperature through the greenhouse effect. Results from climate models suggest that a doubling of the CO_2 concentration would give an increase in the global average earth surface temperature of 1.5 – 3°C. Although the consequences of such an increase are not accurately known, it is believed that it could have far reaching impacts on natural ecosystems as well as human activities (El-Hinnawi, 1981).

Trees are a renewable resource, provided appropriate management and conservation measures are taken. The logical immediate response to the fuelwood shortage is to manage the forests appropriately and to plant more trees in plantations, farms, along roads, in shelter belts, and on unused land. Natural forests grow on the average between 1 m^3 and 2 m^3/ha/y, whereas managed plantations grow at least 10 m^3/ha/y, thus yielding about five to ten times more volume per unit area per year (Openshaw, 1980). Therefore, by growing trees in plantations and by appropriate management (using only the annual increment from such plantations), an adequate supply of wood could be sustained. Managed natural and plantation forests covering an area one-tenth of the world's surface (1,300 million ha) could produce an annual increment of 13,000 million m^3 more than needed to meet anticipated future requirements of wood by the turn of the century (Openshaw, 1978).

Several tropical trees are especially promising candidates for agroforestry projects. The leucaena tree (*Leucaena leucoce-phala*) has received a great deal of attention recently as a source of fuel and food. Leucaena – known also as the Hawaiian giant, koa haole, or ipil-ipil, depending on the variety and its location – is a native of Mexico and is one of the world's fastest-growing trees. It can grow 20 m tall within six years. A leucaena plantation can annually provide up to 50 tonne of wood per hectare, five times the average of cultivated pines in temperate regions. Because of its spectacular growth potential, utility companies in Hawaii and the Philippines are planting leucaena as a fuel for some of their power plants. By early 1982, the National Electrification Company of the Philippines plans to establish 10,000 hectares of the tree to power a three-megawatt generating facility (the area is suited to provide wood for a 75 MW power plant; Semana, 1979). A Japanese company in the same country is planting 5,000 hectares of leucaena and plans to convert the wood into charcoal for use in the steel industry. Leucaena, a member of the legume family is able to replenish the soil with nitrogen through the action of bacteria in root nodules and is thus particularly well suited to agroforestry schemes. In several countries of Southeast Asia, it provides shade for coffee and cacao groves. In northern Australia, leucaena is intercropped with pangola grass, and both plants provide nutritious fodder for cattle (Smith, 1981).

Among the many other trees now being used in reforestation schemes in the tropics, red kaliandra also deserves special mention. A native of Central America, the bush can reach 3.5 m in just nine months. Red kaliandra easily produces up to 33 tonne of wood per hectare every year (Silitonga, 1979) and is often planted with food and ornamental plants in the vicinity of homes. Because of its fast growth, kaliandra competes with, and eventually suppresses, lalang, a tough perennial bunch grass that often invades and takes over cutover areas of Southeast Asia. In Indonesia, 34,000 hectares of red kaliandra have been established as a buffer zone around national forests to protect trees from fuelwood gatherers.

Eucalytpus is the tree planted most widely throughout the world for fierwood production. Several of the hundreds of species have adapted to a wide range of environments, from the cool highlands of the Andes to the warm, moist equatorial lowlands of Amazonia. Its ability to withstand drought, to regenerate profusely, and to grow luxuriantly within a short period are major reasons for its widespread dispersal. In Brazil, where yields average 12 tonne per hectare annually, eucalyptus is cultivated for charcoal and for methanol production. The area of eucalyptus is likely to expand dramatically in the next few years in Brazil and elsewhere. To supply a plant with sufficient feedstock to manufacture 1,000 tonnes of methanol a day, for example, 72,000 hectares of eucalyptus must be harvested, on a seven-year cutting cycle (Smith, 1981).

Several problems are, however, inherent to such energy plantations: land areas required (where land is scarce and there is need for food production), fertilizer requirements, water requirements (in some areas), pressure on soil productivity, etc. The continuous removal of trees for fuel can exhaust the soils they depend on. Loss of nutrients will be increased* (phosphorous, potash, calcium, nitrogen) and increased amounts of fertilizers would be needed (Smith, 1981). Disturbance of the natural habitat for plant and animal species will also occur. However, through appropriate management of such plantations wildlife can be preserved. On the other hand, energy plantations could reduce the rate of CO_2 increase in the atmosphere, only if the fuel crops are managed for sustained yield (increment).

Energy plantations, especially in heavily cultivated areas, face, however, considerable political and cultural resistance, and the experiences of the tree-planting programmes of many governments in wood-short countries over the past few decades have not always been encouraging. Nevertheless, if the energy needs of developing country villages are to be satisfied without continuing an increased environmental degradation, some way must be found of growing fuelwood on a sustainable basis in addition to the concerted efforts to harness other renewable sources of energy. One such way would be to plant trees that would provide

* The rate of removal of nutrients from the soil depends on the type of the tree planted, its growth rate and rate of removal, etc.

fodder and food for the farmers as well as fuel. These could be scattered throughout the fields or in blocks. If such species as legumes were planted then they would improve the soil at the same time by fixing atmospheric nitrogen.

Fuelwood Combustion

The direct combustion of fuelwood is not without potential hazards. Besides home fires and burns, the combustion of fuelwood results in emissions consisting mainly of particulates, condensible organic compounds, carbon monoxide and polycyclic organic matter (POM). Such emissions could cause respiratory diseases and cancer. In a recent study by the U.S. EPA (1980) it was found that smoke from wood-burning stoves and fireplaces contained 17 priority pollutants, 14 known carcinogens, 4 co-carcinogens, and 6 cilia-toxic agents. The EPA study found that emissions of carbon monoxide and POM are much higher for air-tight woodstoves than for fireplaces; the emission rates in grammes of pollutants emitted per kilogramme of wood burned, are as follows:

	Particulate Matter	Condensible Organics	NO_X	CO	POM
Open Fireplace	2.4	6.7	1.9	22	0.031
Air-tight Stove	4.1	6.0	0.6	186	0.29

The U.S. National Academy of Sciences reported benzo (a) pyrene (BaP) emissions from wood combustion 50 times higher than those from oil combustion (Morris, 1980) as follows:

	ug BaP/10^6kJ
Oil	850
Wood combustion	47,000 (NAS)
Wood Fireplaces	42,000 (EPA, 1980)
Wood Stoves (air-tight)	127,000 (EPA, 1980)

Although it is only recently that data have become available on emissions from wood stoves (Butcher and Buckley, 1977; Butcher and Sorrenson, 1979; EPA, 1980), the available data base and overall understanding is much less than might be desired considering the potential impact of emissions from wood stoves. However, in an attempt to evaluate the health risks of wood combustion, Morris (1980) pointed out that particulate emissions from wood combustion are at least 50 times higher in their potentially carcinogenic activity than particulate emissions from oil burners, which emit about 850 ug BaP/10^6 kJ*.

* Morris (1980), on the basis of USA practices in wood harvesting, estimated the following occupational accident rates:

	Effects per 10^{12} kJ Whole Tree Removal	Effects per 10^{12} kJ Residue Removal
Deaths	0.10	0.65
Injuries	15.9	100
Lost work days	149	926

Residue removal is more hazardous since the material is more dispersed and more labour must be expended per unit of energy recovered.

Another important environmental aspect of fuelwood combustion is the great amount of energy wasted. Cooking on an open fire requires more than five times as much energy as on a kerosene stove (Eckholm, 1975). The commonly used cooking stoves in the developing countries have efficiencies of between 6 to 10% (i.e. from 90 to 94% of the total heat value of the fuelwood is wasted). Several attempts have been made to increase the efficiency of wood stoves. In Indonesia, the Singer type stove with an efficiency of 27% led to substantial savings in wood comsumption. Other types of efficient stoves include the Indian Junagadh stove (30% efficiency), the Guatemala Lorena type (15–20% efficiency) and the New Nepali Chulo (about 20% efficiency). On improving stove efficiency, it should be taken into account that fire making in rural households is not intended only for cooking, but also for other purposes, for example, for drying of different products. Improving the cooking efficiency of the stoves should not be at the detriment of these and other functions (Wiersum, 1979). It should be also noted that the successful introduction of simple wood burning stoves depends not only upon choice of technology and technical efficiency but also upon another basic consideration: the needs and preferences of stove users. An improved stove which is not culturally and socially acceptable will not be used. On the other hand, the findings in the U.S. EPA study, referred to above, that more efficient stoves tend to emit more POM and carbon monoxide emissions (the stove acts more like a wood gasifier or pyrolyzer, and produces increased amounts of these pollutants) call for a detailed assessment of the health implications of efficient stoves before they can be put into wide-scale application.

Charcoal Production and Use

When wood or other ligocellulosic biomass (e.g. a wide variety of agricultural residues and wastes like cotton gin waste, wood wastes, peanut hulls, etc. and municipal solid wastes) is heated in the absence of air it breaks down both physically and chemically into a complex mixture of liquids and gases and a residual char, commonly known as charcoal. This process, known as "pyrolysis" can be carried out under a wide variety of conditions, the two extremes being either to maximize the production of charcoal, or to convert the total organic portion of the biomass into combustible liquids and gases (liquefaction and gasification). Pyrolysis is not a new technology. Charcoal has been produced in primitive earthern kilns from the dawn of civilization. But, using such primitive techniques, valuable by-products (oil/tar and gas) have been wasted. Modern pyrolytic conversion processes provide a means for recovery of these materials. For various reasons, these processes have been primarily applied to municipal refuse and agricultural and forestry process wastes. Thus, pyrolysis in its modern form offers a means for producing clean-burning zero-sulphur, storable and transportable fuels from otherwise wasted materials which are themselves major sources of pollution.

The bulk of the world's charcoal is produced from wood by technologically primitive means. Traditional earth kiln methods yield one tonne of charcoal from 12 m^3 of wood (8.6 tonne air-dry wood). Modern industrial processes in commercial use are capable of producing charcoal from wood at yields close to the theoretical one which is 3.3 tonne of wood per tonne of charcoal. Thus they are 2 to 3 times more efficient in their use of the wood than the traditional methods. A more common conversion factor is 6 m^3 (or 4.3 tonne air dry wood) of roundwood per tonne of charcoal (Openshaw, 1978).

World charcoal production has been estimated to be about 4.6×10^6 tonnes in 1978 (United Nations, 1979). However, it should be noted that this figure is based on information supplied by a limited number of countries; many countries do not have adequate information on charcoal production. In many developing countries much charcoal manufacture is carried out by itinerants working for only part of the year and keeping no records of what they produce and sell. Even in developed countries, the statistics of charcoal production are not always readily available and are often published in an ad hoc fashion. Openshaw (1978) estimated that about 570 million m^3 of roundwood were used in 1976 for charcoal production. Assuming that the bulk of charcoal has been produced by traditional earth kilns, which use about 12 m^3 to produce one tonne of charcoal, the production of charcoal in that year would be about 48 million tonne, which is nearly 10 times the official estimate.

Charcoal offers a number of advantages over wood as a fuel. Charcoal is easier to transport, store and distribute, more efficient in burning, less polluting, and it has special advantages in some industrial uses. Charcoal has a higher calorific value than fuelwood. Comparing them both on a dry basis ready to use, charcoal has a calorific value 2–3 times that of wood (the average calorific value of charcoal is 30,400 kJ/kg while that of air dry wood is about 14,700 kJ/kg). Traditional charcoal making produces 1 tonne of charcoal from about 8 tonne of wood; the net efficiency of using that tonne of charcoal as fuel is nearly equal to that of using the 8 tonnes of wood (wood stoves are about 6–8% efficient while charcoal appliances are 27–35% efficient). If modern conversion techniques are applied, less wood would be required to produce 1 tonne of charcoal and the net efficiency of the charcoal system will be much higher. In addition, appropriate techniques make it possible to transform into charcoal large volumes of wood of all dimensions which would otherwise have been destroyed in land clearing or left on the ground in exploited areas. Charcoal is traditionally preferred for a number of domestic uses as well as industrial, mainly metallurgical and chemical.*

The environmental impacts of production of charcoal are essentially the same as those described above under fuelwood, from which the bulk of charcoal is obtained. However, the use of charcoal is less polluting than fuelwood. It is smokeless, an essential requirement for cooking and heating in closed places, with much less emissions of particulates and hydrocarbons. Without adequate ventilation carbon monoxide poisoning may, however, occur.

* In Brazil, for example, charcoal supplies 40% of the energy needs of the steel industry. In Uganda, the fuel is used by cement manufacturers.

Modern Pyrolysis

A wide variety of modern pyrolytic processes have been recently developed and applied to a variety of biomass resources: wood, agricultural and agro-industrial residues, urban refuse, etc. These processes aim at maximizing the conversion of this material into charcoal, oil/tar and/or gases (for detailed description of pyrolytic processes, see Weinstein and Toro, 1976; Tatom *et al.*, 1975; Golneke and McGauhey, 1976; Anderson and Tillman, 1977; Tillman *et al.*, 1977; Jones, 1978; Johns *et al.*, 1978).

The primary variables regulating the amount and quality of product yields are reaction time, reaction temperature, pressure, the presence of catalysts, the reactor type, etc. Considering the wide variety of processes developed or in development, the choice of a system for a particular application presents a difficult problem and involves many factors, including the type of feed to be utilized, its proximity to the conversion equipment, the desired product yields and other site-specific requirements. For example, a unit to convert agricultural wastes might emphasize char and oil production since a gas would likely be difficult to utilize. On the other hand, a system applicable to urban wastes might favour gas and/or oil production, since equipment to use these fuels may already be in use in many homes and factories. For rural applications where charcoal, oil/tar are suitable products, a low to medium temperature, vertical bed reactor using dried feed is most appropriate, while for urban uses where gas is more suitable, a high temperature convertor or a "flash pyrolysis" system would perhaps be most desirable.

Besides their use as industrial fuels, the products of pyrolysis have numerous applications. Recent US EPA-supported work (Polk, 1977) indicates that the pyrolytic oils contain as many as 14 organic compounds, several of which are valuable compounds. Computation indicates that by separating these compounds from the remaining oil which would be used as a fuel, the value of the combined product would be increased by almost 200%. The charcoal produced by pyrolysis, depending on the feed characteristics, the reaction temperature, etc., may be activated, thus allowing uses in water and perhaps air filtration systems. Moreover the charcoal, through densification could provide a useful fuel for mobile gasification plants to power heavy equipment such as tractors, trucks, etc. Pyrolytic gases have been used as fuel for a slightly modified spark ignition engine which operated at about 65% of its rated power (Tatom, 1978).

BIOGAS

Animal dung (cow dung cakes) has been used as fuel for centuries in rural areas of the developing countries. The direct combustion of dung cakes has deprived the land of a good fertilizer. Each tonne of cow dung burned may mean a loss of about 50 kg of food grain. The recent expansion in biogas technology can largely contribute to the alleviation of this problem (El-Hinnawi, 1981). It would, however, be misleading if one would say that the conversion of biomass (organic matter; e.g., human and animal wastes, agricultural and agro-industrial residues, etc.) into biogas (about $\frac{2}{3}$ methane and $\frac{1}{3}$ carbon dioxide) by anaerobic digestion (fermentation by bacteria in an atmosphere devoid of oxygen) is a new technology. Anaerobic digestion has been known for more than three centuries: during the past 50 years or so, some countries in Europe (e.g., the Federal Republic of Germany, France, etc.) and in North America have built anaerobic digesters to treat their sewage and in many cases have used the resulting biogas "sewage gas" as a source of energy. In 1951, for example, 48 sewage treatment plants in the Federal Republic of Germany provided more than 16 million m^3 of biogas, 3.4% of which was utilized for power production, 16.7% for digester heating, 28.5% was delivered into the municipal gas supply system, and 51.4% was converted to vehicle motor fuel (Tietjen, 1975).

Although small biogas plants have been built in some developing countries (especially in the People's Republic of China and India) in the 1950s, the main thrust to develop biogas plants came in the late 1960s and early 1970s. The so-called energy "crisis" of the early 1970s intensified such programmes, which are now being carried out in more than 30 developing countries. Some examples could be given here. In India, more than 36,000 biogas plants (called "gobar" gas) were built up till 1975–76; now the figure is about 80,000, mostly small family-size plants. In the People's Republic of China, there are about 8 million such plants; in Korea about 27,000 and a few hundreds each in Pakistan, Nepal, the Philippines, Thailand, Indonesia, Sri Lanka, etc. Other countries (e.g. Tanzania, Botswana, Ethiopia, Kenya, Egypt, etc.) have installed experimental and/or a small number of operational biogas plants. Although most of the biogas plants constructed are of family size providing from 2–5 m^3 of gas per day (with digester size of about 10–25 m^3), community-size plants have been built in some countries (e.g. in the People's Republic of China, India, Sri Lanka).

Anaerobic Digestion

Anaerobic digestion of biomass is a two-stage biological process and each stage is performed by a distinct group of bacteria. The first stage involves the conversion of the fats, carbohydrates and proteins in the biomass to simple organic acids such as acetic and propionic acid. The "acid-forming" bacteria reproduce rapidly and are not sensitive to changes in their environment. The second stage of the process involves the transformation of the acids to methane and carbon dioxide. In this step a sensitive group of bacteria (the methane-forming bacteria) use the organic acids produced in the first stage as substratum and produce the end products: carbon dioxide, methane and traces of other gases. The methane-forming bacteria are relatively few in number and do not reproduce rapidly. They are extremely sensitive to the presence of oxygen and in general require a special environment to function properly. The acid-forming and the methane-forming bacteria work simultaneously in the anaerobic digestion process. In an anaerobic digester oxygen is excluded and conditions such as temperature, loading rate, etc. are controlled so that the methane-formers are in balance with the acid-formers. Otherwise, the methane-formers would be inhibited and in some cases would stop functioning entirely (see, for example, NAS, 1977; UNITAR, 1976; Maramba, 1978).

The quantity and composition of the gases produced during anaerobic digestion are a function of the fraction of the total waste that is available to the anaerobic bacteria, i.e. the bio-degradable fraction, and the operating environmental conditions of the process. The more bio-degradable the waste material, the greater the quantity of methane generated per quantity of waste added to a digester. Not all wastes are equally bio-degradable and effective in producing methane. The bio-degradable fraction of waste will vary, being a factor of the characteristics of the material, the food ingested to generate the human or animal wastes, and how the wastes were handled prior to digestion. Untreated human wastes or fresh manure are highly bio-degradable material. When exposed to the natural environment for some time, natural degradation of the organic fraction takes place and the bio-degradable fraction will decrease (Loehr, 1978).

One of the most important environmental conditions which influence the rate of biogas production is temperature. Although biogas production is possible throughout the 0 to 60°C range, the methane-formers are severely limited below 20°C and above 55°C. It is generally agreed that two temperature ranges exist for good biogas production; they are known as the mesophilic range (20° to 45°C) and the thermophilic range (above 45°C). The choice of a digester operating temperature is extremely important from both a biological and an economic viewpoint. The rate of biogas production is temperature dependent and digesters operating in the thermophilic range produce biogas more rapidly than those operating in the mesophilic range. More rapid production permits shorter retention times to be used for obtaining a given amount of biogas. Sudden temperature changes also affect the performance of the bacteria in an anaerobic digester. A rapid change in temperature results in erratic biogas production. If the change is too great or too prolonged, production may stop entirely.

Other factors that influence the ability of anaerobic bacteria to produce steady quantities of high quality biogas include pH, which is a measure of acidity and buffer capacity which indicates the ability of a material to resist changes in pH. Animal manure usually has sufficient alkalinity to maintain the pH in an anaerobic digester within the 7.0 to 8.5 range. As a result methane production from digesting animal manure usually is a very stable process. If the balance of activity between the acid-forming bacteria and the methane-forming bacteria is upset, a harmful build-up of acids occurs. Unless there is sufficient alkalinity and buffering capacity to offset the effect of these acids the pH will begin to drop indicating that failure may be occurring. Digester failure occurs when the pH gets too low causing the methane-formers to reduce their activity. This results in an increased CO_2 content in the biogas. A further decrease in pH results in complete cessation of all biogas production including CO_2. Possible reasons for declining pH may be one or more of the following: (*a*) the loading rate is too great; (*b*) temperature fluctuations are too great; (*c*) toxic elements may be present in the incoming manure; and (*d*) a build-up of scum may be preventing gas release. If pH declines rapidly, loading should be stopped immediately and should be restarted slowly only when the pH has stabilized. It is possible to reduce excess acidity in a digester by adding alkaline substances such as sodium carbonate or hydrated lime but this practice should rarely be necessary in properly managed animal digesters.

Like all living organisms, bacteria also require essential nutrients to sustain life. Fortunately, these nutrients (N, P, K, etc.) are present in manure in quantities which are adequate for good biological activity. In certain situations, however, the balance between certain nutrients may need adjustment. The relationship between carbon and nitrogen, for example, is important in anaerobic digestion. If there is insufficient nitrogen to allow the anaerobic bacteria to reproduce, carbon dioxide will be the principal gas produced. On the other hand, if the waste material contains excessive nitrogen, ammonia will increase and will inhibit the action of bacteria and gas production will slow or even cease. For optimum methane production, the C:N ratio should be in the range of 25–30 :1(ESCAP, 1980). If the C:N ratio in the waste material contemplated for methane generation is high (as in the case of crop residues), nitrogen rich manure is added to adjust the ratio. On the other hand, if the C:N ratio is low (as in the case of poultry manure), straw or crop residues are added.

Biogas Composition and Yield

Many designs of biogas plants have been proposed: combined digester-gas holder plants, floating gas holder types and separate gas holder units (Figs. 8.2 to 8.7). The plants are built using a variety of materials: brick, cement, concrete, steel, plastic, etc. The input to the digester could be human or animal wastes, agricultural residues, cultivated biomass (algae, water hyacinth, etc.) but a mixture of one or more of these is generally preferable to maintain a carbon/nitrogen ratio of about 25–30:1 which is optimum for biogas production. The biogas produced vary in composition according to the design of the digester and the composition of the feed material, but it is generally in the range of:

Methane:	55 – 65%
Carbon dioxide:	35 – 45%
Nitrogen:	0 – 3%
Hydrogen:	0 – 1%
Oxygen:	0 – 1%
Hydrogen Sulphide:	0 – 1%

The calorific value of biogas varies with the methane content, and is generally in the range of 20–26 MJ/m^3, due to the presence of CO_2 (pure methane has an energy content of 37.1 MJ/m³). Some biogas plants have been provided with means to scrub off the carbon dioxide from the biogas to raise its calorific value.

The amounts of biogas produced depend on the conditions of anaerobic digestion and on the nature of the material digested (Zelter, 1979). Several estimates have been given in the literature for the biogas output. However, it is difficult to compare these estimates since in many cases the conditions under which the yields are given are not specified.

Fig. 8.2 Fixed Dome Digester (China)

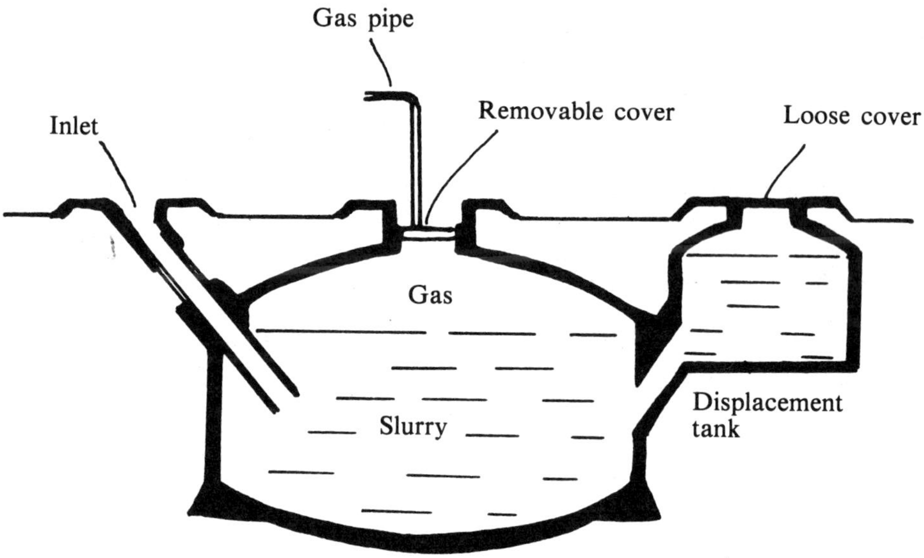

Fig. 8.3 Chinese Biogas Plant

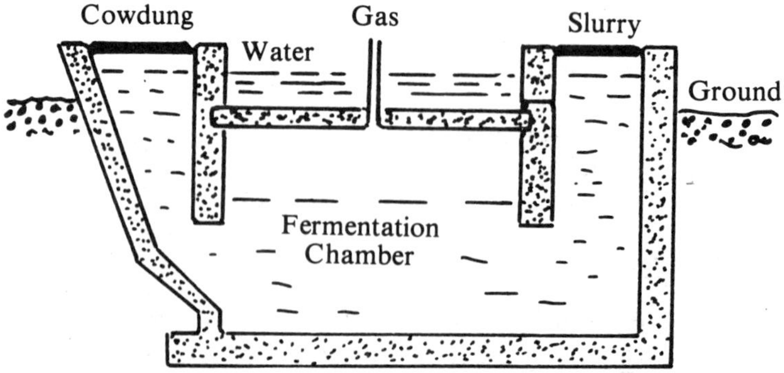

Fig. 8.4 Biogas Plant with Steel Gas Holder

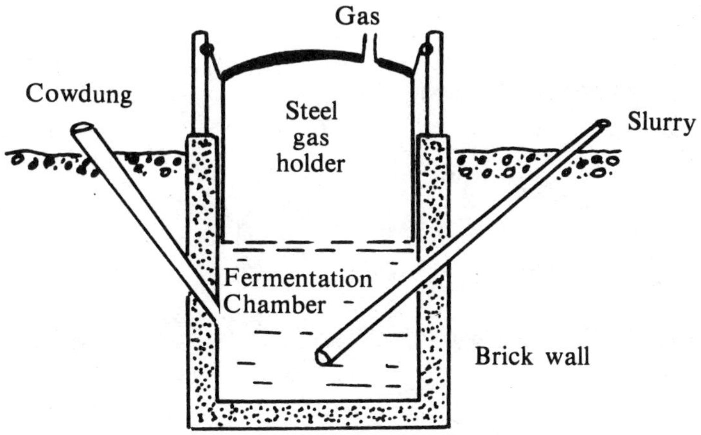

Fig. 8.5 "Sausage" Bag Digester

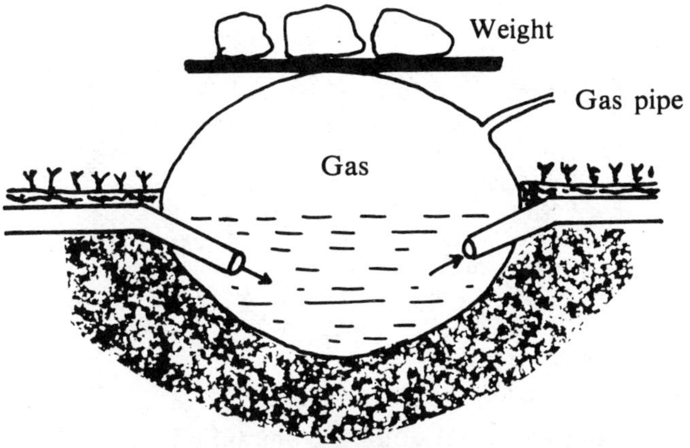

Fig. 8.6 Two-Chamber Digester with Floating Gas Holder

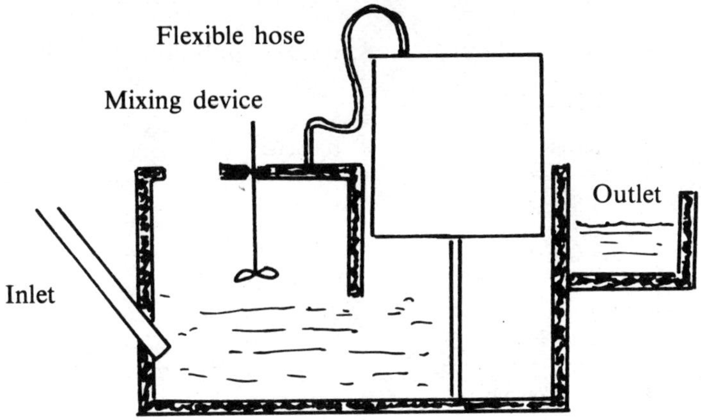

Fig. 8.7 Jar Digester with seperate Gas Holder

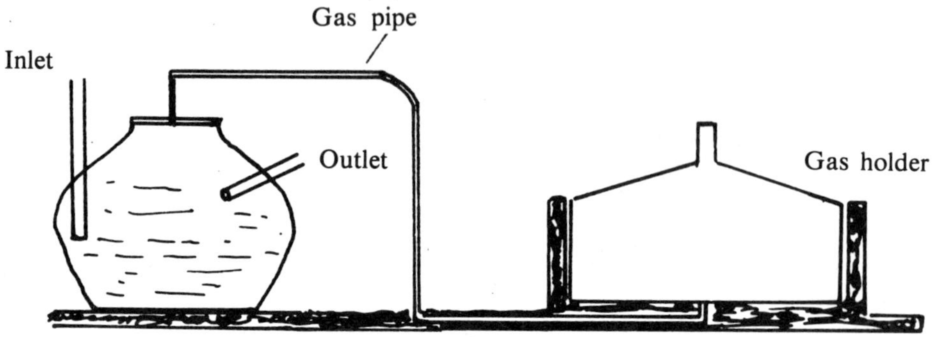

Table 8.2 gives the biogas yields possible per unit mass of *fresh* matter (wet) fed to the digester and completely destroyed.

Table 8.2 Biogas Yield per Unit Mass of Fresh Matter Completely Digested

Material	Biogas yield (m^3/tonne)
Cattle dung	22– 40
Pig manure	40– 60
Poultry	60–115
Human faeces	20– 28
Crop residues	30– 40
Water hyacinth	40– 50

After ESCAP (1980)

Table 8.3 gives the estimated domestic animal population of the world and Table 8.4 gives the estimated annual production of dung and biogas yield.

Table 8.3 Representative Domestic Animal Populations (in million head)

Animal	World Total	Developing Countries
Cattle	1,201	685
Buffalo	132	99
Horses, Mules	121	73
Camels	14	13
Pigs	674	120
Sheep, Goats	1,447	745

After FAO (1975).

Table 8.4 Estimated Annual Production of Dung and its Energy Content

Source	Fresh manure* production per head kg/y	Calculated** amount of manure t/y	Estimated*** amount of biogas production m^3/y	Energy content**** MJ
Cow	5.400	64.9×10^8	19.5×10^{10}	42×10^{11}
Buffalo	7,000	9.2×10^8	2.8×10^{10}	6.2×10^{11}
Horses, Mules	1,700	3.3×10^8	0.66×10^{10}	1.5×10^{11}
Pigs	1,300	10.1×10^8	55.1×10^{10}	11.2×10^{11}
Sheep, Goats	500	7.2×10^8	3.6×10^{10}	7.9×10^{11}

* Fresh dung production (ESCAP, 1980); the amount of manure produced varies according to the weight of the animal, amount and nature of animal feed. The values given here are average values nearly representative to conditions in developing countries in general.
** Calculated figures using data in Table 8.3.
***Estimated figures using data in Table 8.2.
**** Calculated figures assuming an average calorific value of 22 MJ/m^3 of biogas.

Use of Biogas

Many options exist for utilizing biogas (Fig. 8.8). It can be used directly in gas-burning appliances for cooking, lighting and refrigeration or it can be used as fuel for internal combustion engines. In the latter case, the hydrogen sulphide content of the biogas must be maintained below 1.5 g/m^3 to avoid corrosion. The H_2S contained in biogas from manure is sometimes greater than this limit and in such cases should be removed to prevent damage to the engine. Biogas is extensively used in India, the People's Republic of China, Korea, Nepal, and several other countries for cooking and lighting. It is also used for running irrigation pumps and machinery and for electricity production. For example, biogas from sewage in Fushan (Guangzhou, China) is used to operate a power plant with a capacity of 630 kW (FAO, 1979).

By-Products of Anaerobic Digestion:

The effluent and sludge remaining, after digestion has taken place, is a rich and effective fertilizer. Anaerobic digestion of plant residues and animal wastes conserves nutrients needed for the continued production of crops; the only materials removed form the system for purposes other than putting back on the land are the gases generated – methane, carbon dioxide and hydrogen sulphide. The fate of nitrogen is particularly important; a major advantage of anaerobic digestion of biomass is the conservation in organic or ammonium nitrogen forms, of practically all the nitrogen present in the material used. The sludge produced by anaerobic digestion is believed to have a fertilizer value greater than that of the original raw material (perhaps due to more concentration in relation to volume of material). It has no offensive odour when spread on land and rodents and flies are not attracted to the remaining solid or liquid residues. It is claimed that the use of biogas slurry as a fertilizer had led to an increase in the yield of some crops by 10 to 28% in the People's Republic of China and India as compared to the use of excreta (Van Buren, 1979). The use of biogas slurry can also give protection to crops during prolonged drought and also reduce the weeding cost as fertility of weeds is destroyed in the fermentation process. Experiments in the Philippines have shown that rice fertilized with commercial urea had an average yield of 6.5 tonne/ha whereas rice fertilized with manure sludge produced 8.3 tonne/ha (Eusebio and Rabino, 1978). The effluents can be used also for growing algae that can be used as animal feed and input material to the digester; the water from the algae ponds can be used to feed fish ponds. Yields of *Tilapia* of over 7 tonne/ha were achieved in field experiments (other species of fish can also be produced).

Fig. 8.8 Possible Applications of Biogas (After ESCAP, 1980)

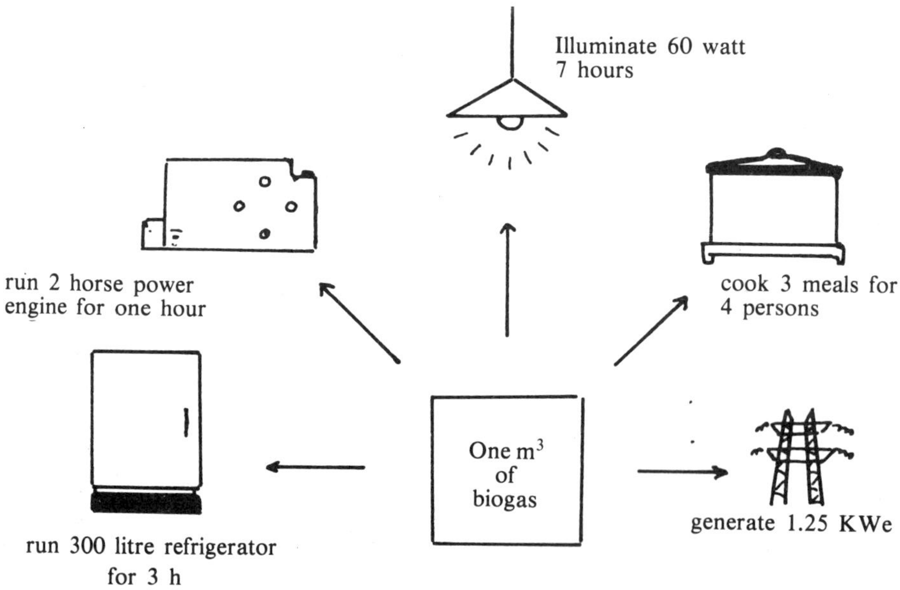

It should, however, be noted that effluents resulting from anaerobic digestion cannot be directly applied as a fertilizer or for fish farming because of their toxicity (largely due to the presence of hydrogen sulphide). Sludge conditioning is, therefore, important to eliminate such toxicity and permit its use for different purposes. At the Maya Farms in the Philippines, sludge-conditioning plants have been built adjacent to the biogas plants (Maramba, 1978). The solids recovered from the sludge have been processed into an excellent animal feed material (dried sludge was found to be rich in vitamin B_{12}, a growth-promoting factor in animal feeds).

Environmental Aspects of Biogas Production and Use

The environmental impacts of production and use of biogas should be considered in the context of "Biomass–Biogas–Bioproductivity" system (El-Hinnawi, 1981). Anaerobic digestion provides a means for environmentally-sound management of organic wastes; or as expressed by the Chinese "turning the harmful into the beneficial". This is of particular importance in rural areas of the developing countries which generally lack adequate systems for the disposal of organic wastes. Such wastes pose the gravest health hazards because they carry the pathogens of the water-borne diseases (e.g., cholera, typhoid, dysentery, etc.) and are a breeding milieu for different parasites (e.g. schistosomes, hookworms, etc.) which are causing widespread incidences of different diseases in rural areas Anaerobic digestion of organic wastes eliminates to a large extent these organisms, leading thereby to improvements in general public health. The use of biogas can also reduce the demand for firewood and charcoal in some areas, leading to the conservation of forest resources and reduction of the process of soil degradation and desertification.

To illustrate the potential of wide scale use of anaerobic digestion, it is estimated that if 60% of the organic wastes of the farm animals in Southeast Asia had been collected in 1975 and digested for biogas production, the energy produced would have been equivalent to about 57 million barrels of oil (Eusebio and Rabino, 1978). This amount would have reduced the total oil imports of Southeast Asia in that year by about 15%. In addition, the amount of effluents that would have been used as fertilizer would have saved some 200 million US dollars worth of chemical fertilizer.

Biogas is a cleaner fuel than fuelwood or coal and its use for cooking in rural areas of the People's Republic of China and other countries has eliminated the pollution and health hazards associated with the use of wood or coal. The use of biogas has also encouraged a decrease in wood cutting, which in turn has reduced deforestation and consequent soil degradation.

Several schemes of the "Biomass–Biogas–Bioproductivity" system have been experimented in some countries on a small scale or on a village level (for example, in China, the Philippines, India). The problems to be encountered in large scale schemes are different from those encountered in small scale (or family-size) plants. These include land and water requirements; collection, storage and handling of manure and other wastes; handling of slurry; distribution systems of biogas, and safety questions in relation to handling large quantities of biogas. Safety should be a major concern related to biogas production. Methane is inflammable and when mixed with air within the range of 5 to 15% by volume, is explosive. Safety regulations for various buildings, electrical and equipment installations should be strictly adhered to during the planning, construction and operation of an anaerobic digestion system. For example, areas where biogas might accumulate if a leak were to occur may require an alarm as well as explosion-proof motors, wiring and lights. Biogas lines leading to the flame in a furnace or to the combustion chamber of an engine must have a flame arrester to prevent backfiring into a biogas storage reservoir or into the digester itself. The entire anaerobic digestion system should be checked on a regular basis for leaks and possible safety hazards.

Kuo and Jones (1978) have studied the environmental impacts of large scale production of biogas. Assuming that biogas will be used for electricity generation, Table 8.5 gives the material balances for production of 100 MW(e)y, as calculated from the date given by Kuo and Jones (1978). The water quantities required for digestion are zero when fresh manure is used and waste water is recycled. In case of no recycling, evaporation ponds may be used for liquid disposal to avoid discharge to surface water. Evaporation ponds could produce odours at certain times and become a source of ammonia and H_2S. However, photo-synthetic organisms could oxidize the H_2S and minimize its release. This water effluent might be also used for irrigation. Solid residues from anaerobic digestion of biomass are used as fertilizer and have less environmental impact than undigested manures that are spread directly on agricultural land.

Table 8.5 Material Balances in Production of Biogas to Generate 100 MW(e)y*

Organic waste input (cattle manure, fresh)	1.48×10^6 tonne solid 5.92×10^6 tonne water
Biogas produced	Gross amount: 4.1×10^8 m^3 (8.97×10^{12} kJ Gas used for heating digester: 5.3×10^7 m^3 (1.17×10^{12} kJ Net biogas: 3.5×10^8 m^3 (7.8×10^{12} kJ
Sludge	0.91×10^6 tonne solids 2.77×10^6 tonne water
Wastewater	3.17×10^6 tonne

* Calculated from data given by Kuo and Jones (1978) assuming that 9000 kJ are the minimum required per kWh and that the calorific value of biogas is 22×10^3 kJ/m^3. A part of the biogas produced is used to heat the digester system. Emissions from boilers consist mainly of sulphur and nitrogen oxides; these are calculated as 180 tonne (S) and 163 tonne (N) to produce the amount of biogas necessary for production of 100 MW(e)y.

With the ever-increasing interest in building biogas plants, several basic chemical, microbiological, engineering and social problems have to be carefully studied to ensure the successful wide-scale adoption of anaerobic conversion technology. Several research and development programmes are under way in many countries to study these and other factors that influence the process of bioconversion. Anaerobic conversion should be regarded as an environmentally-sound management technique that protects the environment, increases the bio-productivity and provides an adequate energy source. Such biomass–biogas–bioproductivity integrated systems can lead to substantial and accelerated rural development, increase the base for food production and improve the quality of life for millions of people in rural areas. Biomass–biogas–bioproductivity systems, at the village level making use of human, animal and other organic wastes, can effectively change the gloomy picture of rural areas into a more bright one through the development of rural industries, the provision of local job opportunities and the progressive eradication of hunger and poverty.

POWER ALCOHOL

Methanol

Methanol (CH_3OH) was originally produced on a limited scale in wood distillation plants as a byproduct of the production of charcoal. It has always been regarded as an excellent fuel because it burns clearly and can be readily transported and stored, although it has only about 50% of the calorific value of gasoline on a volume and weight basis. Methanol was used as a fuel during most of the eighteenth and part of the nineteenth century for cooking, heating and lighting, until it was replaced by kerosene and other fuels. Interest in the use of methanol as alternative motor fuel dates back to the early 1910s. Methanol has high heat of vaporization compared to gasoline which results in cooler fuel mixtures and a lower temperature of combustion. These lower temperatures may cause starting problems but they also result in low nitrogen oxide emissions. This is important because these compounds are the major contributors to photochemical smog. Methanol also has a much lower calorific value than gasoline so fuel consumption will be higher on a weight or volume basis and larger gas tanks will be required. However, specific energy consumption (per km) will be lower because of higher compression ratios and the use of simpler emission control systems. Because of its slower burning characteristics methanol is a high octane fuel and therefore should not require aromatic supplements which are normally added to low-lead and lead-free gasoline for improving octane quality. Thus even the partial use of methanol should reduce the carcinogenic risk caused by the presence of these aromatic compounds in the fuel and in the exhaust gases.

One potential problem with gasoline-methanol blends is the possibility of phase separation in the presence of small amounts of water. This could cause problems in the car engines and therefore some method of keeping water out, or some additive to dissolve it in the fuel mixture, would have to be developed. Test data by a number of organizations have indicated that up to 20% methanol can be added to the gasoline without requiring any engine modifications but above this level some minor adjustments to the carburettor and fuel system are necessary (Bene *et al.*, 1978). Another potential problem is toxicity. However, methanol has been used extensively during the past 150 years and there is no medical evidence to indicate that the addition of methanol to gasoline would materially change the hazards associated with the handling of this fuel.

During the 1920s a process was developed for producing "synthetic" methanol by passing a mixture of hydrogen and carbon monoxide over a catalyst at high temperatures and pressures. The hydrogen and carbon monoxide were produced by the reaction of steam with coke; charcoal produced from biomass would have served equally well except for its much higher cost. This process was used until the 1940s when natural gas became available in quantity and the process was then changed from using "water gas" produced from coke to "synthetic gas" produced by the reforming of methane. This is easily the most economical and most direct process known at the present time and all of the current production of methanol is based on the use of natural gas as a raw material. However, several recent studies have focussed on the use of forest and municipal wastes for the manufacture of methanol (Bene *et al.*, 1978). The question at present is that of cost, and it is expected that the economics of producing methanol from biomass will become more favourable in the future as the costs of coal and oil increase.

Ethanol

The production of ethanol can be accomplished in at least three ways: (*a*) direct fermentation using naturally available sugars such as sugar-cane; (*b*) indirectly using carbohydrate or starch sources such as cassava; (*c*) in combination with acid hydrolysis or enzymatic hydrolysis of cellulose sources such as wood which produces sugar that can then be fermented. While alcohol production from sugar-cane or cassava are familiar techniques, this is not generally so in the case of cellulose hydrolysis even though several plants using this process have operated in the U.S.S.R. for about 40 years. However, the economies of these processes, up till recently, have not been attractive; and the primary source of ethanol has been through chemical modification of petroleum derived ethylene which was more economical prior to the recent rapid increase in oil prices. In 1974, 75% of all industrial alcohol produced in the U.S.A. came from this source (Bene *et al.*, 1978). Apart from the more favourable economics it has been more logical to use petroleum as a source of alcohol rather than cereal grains and molasses, i.e. starch and sugars, because these materials have a more important use as a source of food for both humans and animals. However, the cost gap between these two methods will diminish as the supply of petroleum dwindles and the price increases, and, for these and other reasons, there will be a tendency to revert back to the fermentation technology.

Sugar-cane is an excellent raw material for the manufacture of ethanol; sweet sorghum, corn and cassava have also been used. The composition of sugar-cane is about:

		average wt. %
Sugars	Sucrose	14
	Hexoses	1
Fiber (bagasse)		12
Humidity		73

The fermentation of sugars proceeds through the following reactions:

$$\underset{180\text{g}}{\underset{C_6H_{12}O_6}{\underset{(\text{glucose + fructose})}{\text{Hexoses}}}} \xrightarrow{\text{Fermentation}} \underset{92\text{g}}{\underset{2C_2H_5OH}{\text{Ethanol}}} + \underset{2CO_2}{\text{Carbon Dioxide}}$$

$$\underset{342\text{g}}{\underset{C_{12}H_{22}O_{11}}{\text{Sucrose} + H_2O}} \longrightarrow \underset{180\text{g}}{\underset{C_6H_{12}O_6}{\text{Glucose}}} + \underset{180\text{g}}{\underset{C_6H_{12}O_6}{\text{Fructose}}} \xrightarrow{\text{Fermentation}} \underset{184\text{g}}{\underset{4C_2H_5OH + 4CO_2}{\text{Ethanol}}}$$

The production of ethanol from corn proceeds according to the same reaction; the starch should first be converted into glucose through enzymatic conversion:

$$\underset{100\text{g}}{\text{Starch}} \xrightarrow[\text{conversation}]{\text{Enzymatic}} \underset{130.7\text{g}}{\text{Glucose}} \xrightarrow{\text{Fermentation}} \underset{66.4\text{g}}{\text{Ethanol}}$$

Sugar-cane is the most efficient biomass that collects and stores solar energy in a useful form. In areas where the rate of photo-synthesis and carbon fixation are high (e.g. along the Amazon Basin, Equatorial Africa and in Equatorial

Southeast Asia), the yield of sugar-cane is as high as 60 tonnes per hectare per year. Table 8.6 gives the alcohol yield of some selected crops, from which it appears that on a weight basis (litre per tonne of crop), sugar-cane has a higher yield per hectare than other crops and hence the alcohol yield per hectare of sugar-cane is the highest (Jackson, 1976).

Table 8.6 Alcohol Yield of Selected Crops

Crop	Crop yield ha/year (tonne)	Alcohol yield per ha litre	Alcohol yield l/t
Sugar-cane (Brazil)	54.2	3,600	67.0
Sweet Sorghum (US)	46.5	3,554	76.0
Cassava (Brazil)	11.9	2,137	180.0
Corn (U.S.A.)	5.7	2,200	386.0

After Brown (1980).

Several pilot-scale studies are underway in several countries for the production of alcohol from biomass; production programmes have already started in some others, the largest so far is the Brazilian National Alcohol Programme (Yang and Trindade, 1978; CTP, 1979; Trinidade, 1980). "Gasohol" – gasoline and alcohol full blends – have been used in automobiles for decades in Brazil. The earliest recorded experiments with alcohols and gasohols date back to the 1920s. Compulsory alcohol addition to gasoline to a level of 5% was legislated in 1931 to help stabilize the important sugar industry. Excess cane molasses and sugar were converted to alcohol in distilleries attached to sugar mills. The alcohol content in gasohol, however, oscillated considerably depending on the international sugar market. Pioneering experience with neat alcohol (96°GL or 93.8-wt%) fuels in automobiles were performed in the 1920s. Starting in 1974 the performance of Brazilian engines fueled with gasohols containing varying amounts of alcohol has been investigated. The overall conclusion was that all existing Brazilian Otto-engine vehicles, when fueled by gasohol containing up to 20% by volume of alcohol, did not lose performance compared with straight gasoline operation. Furthermore, no engine modifications were required.

Trinidade (1980) summarizes the landmarks of the Brazilian alcohol programme as follows:

1975	Programme launched		
	Production (1974)	:	$0.6 \times 10^6 m^3$
	Alcohol in gasohol	:	$(0.2 \times 10^6 m^3)$
	Gasoline displacement	:	1%

1977	Production	: $1.4 \times 10^6 m^3$
	Alcohol in gasohol	: $(0.6 \times 10^6 m^3)$
	Gasoline displacement	: 5%
	Sao Paulo State gasohol reaches 20% alcohol tests with neat alcohol powered cars	
1978	Production	: $2.4 \times 10^6 m^3$
	Alcohol in gasohol	: $(1.4 \times 10^6 m^3)$
	Gasoline displacement	: 11%
	50 Gasohol blending centres Brazil wide	
1979	Production	: $3.7 \times 10^6 m^3$
	Gasoline displacement	: 14%
	4,000 converted fleet vehicles on neat alcohol	
	Neat alcohol sold in gas stations	
1980	Production forecast	: $4.0 \times 10^6 m^3$
	Gasoline displacement	: Larger than 20%
	200,000 new and converted neat alcohol cars available to individual consumers	
1985	Production forecast	: $10.7 \times 10^6 m^3$
1987	Production forecast	: $14.0 \times 10^6 m^3$

Environmental Impacts of Ethanol Production and Use:

Fermentation alcohol production generates large amounts of effluents which are the raw material washwaters, stillage and flegmass. Stillage is the most salient of these effluents and presents the greatest problem for the environment if inadequately treated. Characterized as the bottom slop of waste residue drawn off the bottom of the first distillation column, stillage is generated at a rate 12 to 13 times greater than the primary product, alcohol. A quick estimate shows that the 7.5 million m^3 of ethyl alcohol forecasted for production in 1985 will generate roughly 94 million m^3 of stillage If not suitably recovered or treated, that volume could have a serious negative impact on the environment. On the average, the pollution potential in terms of BOD in two litres of untreated stillage discharged into inland or coastal waters equals the sewage produced by one person per day. This rate provides an idea of the dimensions of the challenge inherent in the predicted volume of stillage, roughly equal to the sanitary waste (in BOD terms only since stillage contains no pathogenic micro-organisms) of 44 million inhabitants concentrated in Brazil's alcohol-producing regions (Trindade, 1980).

Since stillage normally does not contain pathogenic organisms or toxic compounds, recovery of its minerals and organics is a potentially attractive undertaking. It is technically feasible to convert stillage into marketable products such as fertilizers (Costa Ribeiro and Branco, 1980) and feed additive or into methane as a supplementary energy source (Branco and Costa Ribeiro, 1979).

Economics will ultimately determine the choice of stillage treatment process and the product form. Currently there are two stillage treatment processes being used in Brazil. The most widespread practice is lagooning. The other process takes advantage of the fertilizer value of stillage. Raw stillage is sprayed over the sugar-cane fields, thereby reducing or eliminating the mineral fertilizer requirements (Yang and Trindade, 1978; CTP, 1979).

The energy requirements to produce ethyl alcohol from sugar-cane, cassava and sweet sorghum were studied in detail by Da Silva *et al.* (1978). It has been found that sugar-cane is the more efficient crop for ethyl alcohol production, followed by sweet sorghum and cassava from a net energy viewpoint. Sugar-cane shows a large net gain per year of 89.4×10^6 kJ/ha, which is 1.43 times the total energy consumed in the alcohol system production (agriculture, industry, etc.). Further studies on the energy balance of ethanol production were carried out by Chambers *et al.* (1979) and by Hopkinson and Day (1980). It has generally been found that the evaluation of the energy balance is site-specific and depends on the agricultural and industrial technologies applied; the energy balance should, therefore, be assessed for each case in full detail before embarking on large production programmes.

Up to 20% ethanol can be blended with gasoline without any changes in present-day internal combustion engines. Minor modifications, however, are required for engines running on 100% alcohol. Although ethanol has a lower calorific value than gasoline, ethanol has a higher density and a motor running on ethanol is 18% more powerful than a motor running on gasoline. When added to gasoline, alcohol increases the value of the octane rating. Instead of simply selling gasohol (90% gasoline and 10% alcohol) as a higher grade fuel, refineries could take advantage of this effect by producing a less refined gasoline to combine with alcohol. This would save oil at the refinery (Brown, 1980). Alcohol has the marked advantage of eliminating the use of lead anti-knock additives and of reducing the amounts of hydrocarbons and carbon monoxide emissions. However, the amounts of aldehydes are increased (Table 8.7). The health consequences of such changes in the composition of the emissions have not been studied in detail.

Table 8.7 Compositions of Emissions from Brazillian Otto Engine Vehicles Fueled with Gasoline and Gasohol

Emissions	Emission Rate (g/km)		
	Gasoline	Gasohol	
		10% v	20% v Alcohol
Hydrocarbons	5.0	4.0	3.3
NO_x	0.9	1.1	1.4
CO	75.0	61.0	48.0
Aldehydes	0.03	0.06	0.08

After Hespanhol (1979).

Recently, concern has been voiced that energy plantations (for alcohol production or other sources of energy) may divery agricultural resources away from food production (Brown, 1980). Trinidade (1980) mentioned that in Brazil incentives provided to ethanol production at a time of low sugar prices (prior to 1980) have induced, to a minor degree, expansion of sugar-cane agriculture at the expense of food cropland. However, the situation varies considerably from one country to another according to the availability of land, water, food production and energy situation. Ferchak and Pye (1981), for example, pointed out that with relatively minor adjustments in the agricultural sector, large additional amounts of starch derived from feed corn, surplus and distressed grain, and setaside land could presently be used for ethanol production in the U.S.A., with no evident impact on food production. Utilization of novel cellulose conversion technology can provide fermentable sugars from municipal wastes, agricultural and forest wastes, and ultimately, highly productive silvicultural operations. No conflict with food production would be evident.

References

Anderson, L. and D. Tillman (1977): Fuels from waste. Academic Press. New York.

Bene, J. G. *et al.* (1978): Energy from Biomass from Developing Countries. IDRC Report, Ottawa.

Benemann, J. R. (1978): Biofuels, a survey. Electric Power Research Institute. ER-746-SR Special Report. Palo Alto, California.

Branco, J. R. C. and C. Costa Ribeiro (1979): Anaerobic Fermentation of Stillage. 1st International Symposium on Anaerobic Digestion. Cardiff, Wales. Sept. 17–21, 1979.

Brown, L. R. (1980): Food or Fuel; Worldwatch paper 35. Worldwatch Institute, Washington, D.C.

Butcher, S. S. and D. J. Buckley (1977): Journal of the Air Pollution Control Association 27, 346.

Butcher, S. S. and E. M. Sorenson (1979): Journal of the Air Pollution Control Association 29, 724.

CEQ (1980): Global–2000. Council for Environmental Quality, Pergamon Press, Oxford.

Chambers, R. S. *et al.* (1979): Gasohol: Does It or Doesn't It Produce Positive Net Energy? Science 206, 789.

Chartier, P. (1979): European Community Biomass Programme Conference, Biomass for Energy, London, June.

Costa Ribeiro, C. and J. R. C. Branco (1980): The Use of Stillage in the Ferti-Irrigation of Sugar-Cane Crop Land Dedicated to Fuel Alcohol Production. 2nd International Symposium on Waste Treatment and Utilization. Univ. Waterloo, Ontario, Canada.

CTP (1979): Alcohol: the Brazilian way out of the Energy Crisis: Centro de Technologia Promon. Contribution to UNITAR Conference on Long-Term Energy Resources, Paper CF7/VII/5, UNITAR, UN, New York.

Da Silva, J. G. *et al.* (1978): Energy Balance for Ethyl Alcohol Production from crops. Science, 201, 903.

Earl, D. E. (1975): Forest Energy and Economic Development Clarendon Press, Oxford.

Eckholm, E. (1975): The Other Energy Crisis: Firewood. Paper No. 1. Worldwatch Institute, Washington, D.C.

El-Hinnawi, E. E. (1981): The Environmental Impacts of Production and Use of Energy. Tycooly International Publishing Ltd., Dublin.

EPA (1980): Preliminary characterization of Emissions form Wood-Fired Residential Combustion Equipment. By D. DeAngelis, *et al.* Report EPA–600/7–80–040, Washington, D.C.

ESCAP (1980): Guidebook on Biogas Development. Energy Resources Development Series No. 21. Economic and Social Commission for Asia and The Pacific, United Nations, Bangkok.

Eusebio, J. and B. Rabino (1978): Recycling System in Integrated Plant and Animal Farming. Compost Science March/April, 24.

FAO (1975): Production Yearbook. FAO, Rome.

FAO (1979): China, azolla propagation and small-scale biogas technology. FAO Soils Bulletin 41, FAO, Rome.

FAO (1980): 1978 Yearbook of Forest Products. FAO, Rome.

Ferchak, J. D. and E. K. Pye (1981): Utilization of Biomass in the U.S.A. For the Production of Ethanol Fuel As A Gasoline Replacement Parts I and II. Solar Energy, 26, 9.

Golueke, G. and P. H. McGauhey (1976): Waste Materials annual Review of Energy Vol. 1. Annual Review Inc. Palo Alto, California.

Hall, D. O. (1979): Solar energy use through biology. Solar Energy, 22, 307.

Hespanhol, I. (1979): Composition of Exhaust gases from Brazilian Otto Engine Vehicles. Energia, 5, Nov.– Dec.

Hinckley, A. K. (1980): Deforestation and CO_2. Proceedings Bioenergy '80 Congress. Atlanta, p. 543.

Hopkinson, C. S. and J. W. Day (1980): Net Energy Analysis of Alcohol Production From Sugar-Cane. Science, 207, 302.

Jackson, E. A. (1976): Brazil's National Alcohol Programme Process Biochemistry, June.

Jones, J. (1978): Converting solid wastes and residues to fuel. Chem. Engin. 2, 87.

Jones, J. L. *et al.* (1978): Pyrolysis, thermal gasification and liquefaction of solid wastes and residues – Worldwide status of Processes. ASME 8th Biennial National Waste Processing Conference, Chicago, Illinois, May.

Kuo, M. C. T. and J. L. Jones (1978): Environmental and Energy output analysis for the Bioconversion of agricultural residues to methane. Symposium on Energy from Biomass, paper No. 10, p. 143. Institute of Gas Technology, Chicago.

Loehr, R. C. (1978): Methane from Human, Animal and Agricultural Wastes. In N. L. Brown: Renewable Resources and Rural Applications in the Developing Countries. AAAS Selected Symposium 6, Washington, D.C.

Maramba, F. D. (1978): Biogas and Waste Recycling: the Philippine Experience. Maya Farms, Liberty Flour Mills, Inc. Manila.

Morris, S. C. (1980): Health Aspects of Wood Fuel Use. Brookhaven National Laboratory, New York.

NAS (1977): Methane Generation from Human, Animal and Agricultural Wastes. U.S. National Academy of Sciences, Washington, D.C.

Openshaw, K. (1978): Woodfuel – a time for Re-assessment. Nat. Resources Forum, 3, 35.

Openshaw, K. (1980): Energy Requirements for Household Cooking in Africa with Existing and Improved Cooking Stoves. Proceedings Bioenergy '80 Congress. Atlanta, p. 256.

Pfeffer, P. (1979): UNITAR Conf. Long-Term Energy Sources, UNITAR, UN, New York.

Pimentel, D. *et al.* (1978): Biological Solar Energy Conversion and U.S. Energy Policy. Bioscience 28, 376.

Polk, M. B. (1977): Development of methods for the stabilization of Pyrolytic oil. Annual Report USEPA Grant R–80440010, University of Atlanta, Atlanta, Georgia.

Revelle, R. (1980): Energy Dilemmas in Asia. Science, 209, 164.

Rodin, L. E. *et al.* (1975): Productivity of the World's Main Ecosystems Proc. 5th General Assembly, Special Committee IBP. National Academy of Sciences, Washington D.C.

Saouma, E. (1978): Speech at 8th World Forestry Congress, Jakarta.

SCOPE (1979): The Global Carbon Cycle. SCOPE 13, John Wiley & Sons, New York.

Semana, J. A. (1979): Energy Plantations for Steam Power Plants. Proc. Workshop on Biogas and other Rural Energy Resources, Suva. ESCAP, Energy Resources Development Series No. 19. ESCAP, UN, Bangkok.

Silitonga, T. (1979): The Possibility of Kaliandra wood as a source of Energy. Proc. Workshop on Biogas and other Rural Energy Resources, Suva. ESCAP, Energy Resources Development Series No. 19. ESCAP, UN, Bangkok.

Smith, N. (1981): Wood: An Ancient Fuel with a New Future. Paper No. 42. Worldwatch Institute, Washington, D.C.

Sørensen, B. (1980): Renewable Energy. Academic Press, London.

Tatom, J. W. *et al.* (1975): A Mobile Pyrolytic System. In W. Jewell Energy, Agriculture, and Waste Management. Ann Arbor Science, Ann Arbor, Michigan.

Tatom, J. W. *et al.* (1978): A prototype Mobile System for pyrolysis of agricultural and/or silvicultural wastes. Georgia Institute of Technology. Final Report. under EPA Grant R–803430. Atlanta, Georgia.

Tietjen, C. (1975): From Biodung to Biogas. In W. Jewell: Energy, Agriculture and Waste Management. Ann Arbor Science, Ann Arbor, Michigan.

Tillman, D. A. *et al.* (1977): Fuels and Energy from Renewable Sources. Academic Press, New York.

Trindade, S. C. (1980): Energy Crops – The Case of Brazil. International Conference on "Energy From Biomass" Session I. Brighton, U.K. Nov. 4–7, 1980.

UNEP (1977): The State of the Environment. United Nations Environment Programme, Nairobi.

UNEP (1978): The State of the Environment. United Nations Environment Programme, Nairobi.

United Nations (1979): World Energy Supplies 1973–78. Statistical Paper J22, United Nations, New York.

UNITAR (1976): Microbial Energy Conversion. UN Institute For Training and Research, New York.

Van Buren, A. (1979): A Chinese Biogas Manual. International Techn. Publ. London.

Weinstein, N. J. and R. F. Toro (1976): Thermal Processing of Municipal Solid Waste for Resource and Energy recovery. Ann Arbor Science, Ann Arbor, Michigan.

Wiersum, K. F. (1970): Firewood in Indonesia. Proc. Workshop on Biogas and other Rural Energy Resources, Suva. ESCAP, Energy Resources Development Series No. 19. ESCAP, UN, Bangkok.

Yang, V. and S. C. Trindade (1978): The Brazilian Gasohol Programme. Symposium on Energy from Biomass and Waste, paper No. 42, p. 815. Institute of Gas Technology, Chicago.

Zelter, S. Z. (1979): Methanogenèse et dechets agricoles. Revue Génie Rural, October.

Zumer-Linder, M. (1976): Unprocessed Wood for Housing, Building and Fencing. OECD/CSTP 1976, OECD, Paris.